Autonomous Agricultural Vehicles

This comprehensive guide to agricultural robots is the ideal companion for any student or professional engineer looking to understand and develop autonomous vehicles to use on the modern farm.

With world hunger one of the modern era's most pressing issues, autonomous agricultural vehicles are a key tool in tackling this problem. Smart farming can increase total factory productivity through designing autonomous vehicles based on specific needs, in addition to implementing smart systems into day-to-day operations. This book provides step-by-step guidance, from the theory behind autonomous vehicles, through to the design process and manufacture. Detailing all components of an autonomous agricultural vehicle, from sensors, controlling algorithms, communication and controlling units, the book covers topics such as artificial intelligence and machine learning. It also includes case studies, and a detailed guide to international policymaking in recent years.

Suitable for students and professionals alike, this book will be a key companion to those interested in agricultural engineering, autonomous vehicles, robotics, and mechatronics, in mechanical, automotive, and electrical engineering.

Autonomous Agricultural Vehicles
Concepts, Principles, Components, and Development Guidelines

Ali Roshanianfard
Assistant Professor, Department of Biosystem Engineering,
University of Mohaghegh Ardabili, Ardabil, Iran

Sina Faizollahzadeh-Ardabili
Ph.D., Department of Biosystem Engineering,
University of Mohaghegh Ardabili, Ardabil, Iran

CRC Press
Taylor & Francis Group
Boca Raton London New York

CRC Press is an imprint of the
Taylor & Francis Group, an **informa** business

Designed cover image: Shutterstock

First edition published 2024
by CRC Press
6000 Broken Sound Parkway NW, Suite 300, Boca Raton, FL 33487-2742

and by CRC Press
4 Park Square, Milton Park, Abingdon, Oxon, OX14 4RN

CRC Press is an imprint of Taylor & Francis Group, LLC

ISBN: 978-1-032-27655-7 (hbk)
ISBN: 978-1-032-28455-2 (pbk)
ISBN: 978-1-003-29689-8 (ebk)

DOI: 10.1201/9781003296898

Typeset in Times
by MPS Limited, Dehradun

Contents

Preface

The book you have in your hand, and you probably will decide to read it, will try to discuss autonomous agricultural vehicles (AAVs) and the various aspects of this futuristic subject, away from the non-original content and unnecessary discussion. We wish that one day the results of all the scientific research in this field will facilitate the agricultural process, reduce the concern of food restrictions, and provide access to healthy food for all humans. When this book was written, the culture of using agricultural robots had not yet found its place. Once there were few smartphones, and it was not common to use them, but now most people enjoy their advantages. We believe that AAVs and all intelligent agricultural systems will find their place and application in the next few decades. Our world is heading toward war, drought, famine, violence, and destruction at an indescribable speed. We wish to overcome many of these problems one day by relying on science and rationality. The land we call earth needs concerned, peace-loving, faithful, and hardworking people. Hoping for a day when our dreams come true and geographical boundaries will not prevent us from promoting peace, friendship, and solving problems.

We are very grateful to all those who helped us in writing this book. Thank you for sharing your necessary corrections and valuable comments with us to improve this book's scientific level.

Ali Roshanianfard and Sina Faizollahzadeh-Ardabili

Author Biographies

Ali Roshanianfard is a smart farming researcher and assistant professor (biosystem engineering, University of Mohaghegh Ardabili) originally from Ardabil, Iran. He received a PhD in agricultural engineering – vehicle robotics from Hokkaido University, Japan. After starting research in futuristic farming, he embarked upon careers as a post-doc researcher (vehicle robotics, Hokkaido University, Japan), head of the scientific center (Precision Farming Research Association, University of Mohaghegh Ardabili, Iran), and vice president (Science and Technology Park of Ardabil Province, Iran). He is currently the reviewer of many top-rated academic journals. His research focused on the application of information technology, sensors, control, cultivation, and processing systems in agriculture. He is collaborating with farming robot laboratories international industries, and startups.

Sina Faizollahzadeh-Ardabili has been a member of the National Elites Foundation in Iran, since 2017, which works on smart systems and artificial intelligence technologies. He received his PhD degree in renewable energies, especially in smart energy systems. His research during his PhD studies has been on the application of artificial intelligence systems and smart systems. He has several highly cited review papers on the application of machine learning and deep learning systems. He is collaborating in the AI lab as a researcher now. He had been among the top 2% of scientists based on the Standford University study since 2021.

1 Introduction

1.1 INTRODUCTION

Since many years ago, it has been an open secret that autonomous vehicles optimize safety, reduce waste, and save money. All types of vehicles, from residential lawnmowers to combine harvesters, from vacuum cleaners to oil-tank cleaning systems, and from delivery drones to jumbo airplanes, will soon be distinguished by these potent benefits. In the near future, the automobile will be shaped by autonomy. But as of now, the technology needed to make these systems a commercial reality is still out of reach [1–3]. Technology has advanced in the agricultural sector, from tractor driver-assistive technologies like RTK-GPS displays to autonomous, self-driving platforms that can do agricultural chores without human interference [4].

Agriculture is changing drastically thanks to automation and robotics. The benefits of agriculture automation are clear: lower costs for customers, a significant reduction in the environmental impact of farming, and effectively lower labor costs overall. Agriculture automation businesses are launching the farming industry into a modern environment with self-driving tractors, weeding robots, and controlled environment agriculture [5–7]. Automated agricultural vehicles can help ease worries about labor intensity, lower input costs, and increase profitability. Since 2000, many advancements have been made in autonomous guiding for agricultural tractors. Creating autonomous vehicles or field robots for agricultural activities is the ultimate goal of more recent research and development efforts to increase autonomy [8].

Over 10 billion people are expected on the planet by 2050, necessitating a sustained rise in agricultural output. Given this, the use of autonomous cars in precision agriculture is one of the critical concerns that should be considered to increase efficiency [9]. There is little doubt that safer designs, vehicle management systems, and significant advancements in personal protective equipment are required when autonomous vehicles and workers on foot are present simultaneously [10]. To meet these immediate and long-term problems, we wrote a thorough book covering autonomous cars in the agriculture industry. An introduction, a description of the components, platforms, mechanical components, sensors, communication, agricultural and controlling logic, controlling units, performance indicators, external attachments for manipulation, and a case study are included in the ten chapters that make up this book.

1.2 GLOBAL AGRICULTURE INDUSTRY

A sustainable and inclusive agricultural industry is needed to meet global development goals. Agricultural development is one of the most effective

DOI: 10.1201/9781003296898-1

strategies for reducing extreme poverty, fostering shared prosperity, and feeding the estimated 9.7 billion people by 2050. Compared to other sectors, the agriculture sector's growth is two to four times more successful at increasing the incomes of the poorest people. The year 2016 revealed that 65% of working poor individuals relied on agriculture for their livelihood [11]. To combat hunger, malnutrition, and poverty in the world's poorest nations, the Global Agriculture and Food Security Program, a multilateral finance mechanism, collaborates with various public and private sector partners. The World Bank serves three capacities: trustee, project execution partner, and secretariat host [11].

In collaboration with UNDP, the UN Food and Agriculture Organization (FAO), the Global Landscapes Forum, and the Food and Land-use Coalition, the World Bank is the platform's lead organization. The Global Environment Facility funds FOLUR. FOLUR seeks to support ecologically sound, integrated landscapes and green the supply chains for eight essential food commodities [11,12]. The inability to expand arable land coverage to improve agri-food production and the fast-growing world population highlights the need to boost industry productivity, particularly in developing nations. Digitalization can accomplish this goal by utilizing cutting-edge gear, the Internet of Things (IoT), and other intelligent technologies while increasing yields and reducing labor and costs [13].

The market is as significant as any other global business, thanks to the USD 2.4 trillion contribution from the global agriculture sector. Over 1 billion people are employed worldwide in the agriculture industry, which the top ten agricultural corporations dominated in the world in 2020. Cargill, which earned USD 114.69 billion in revenue last year, is the largest agriculture firm in the world. ADM and Bayer make up the top three agricultural companies. Population and income are the two main factors influencing food demand, and both are growing. The world's population will increase from 7.4 billion in 2016 to 9.1 billion by 2050 [14].

The 20 largest countries producing agricultural products produce a total of 1,531.4 million tons, of which 34% is produced only in the United States of America. Other countries in this continent accounted for 23% of agricultural product production. The countries of the European continent have 17%, and the countries of the Asian continent have 23% of the production share. The amount of agricultural production in Africa was equal to 67.5 million tons, and five countries from this continent are among the major countries producing agricultural products. The study shows that seven European countries are on the list of large producers of agricultural products, and five Asian countries are on this list.

The production of agricultural products is essential for the world economy because this industry is the key to the world's food supply. With the increase in population, there is a concern that the production of agricultural products will be less than the world's needs. But in addition to population growth, one of the challenges facing the world economy, the global warming change and its negative effect on the efficiency of agricultural production, has also created a big problem for the world. The United Nations wrote in its report: "At present, when climate change has brought significant changes to our world, the agricultural industry is more critical than ever. Climate change is the main factor affecting the global agricultural scenario because it has reduced the efficiency of agricultural production worldwide and increased the need to invest in new technologies in this industry. The lack of water in the world and the increase in the earth's temperature caused the production of many products worldwide to decrease".

This issue caused the FAO to warn about climate change and its effect on the agricultural industry and to ask the world to think of a solution to solve the problem of climate change. In its latest detailed report, the U.S. Department of Agriculture introduced the 20 largest countries producing agricultural products and made this classification based on the amount of production in 2019. According to this report, the American continent and the United States of America have been the significant producers of agricultural outcomes in the world.

In 2019, the United States produced 513.74 million tons of agricultural products, and this industry contributed more than 100 billion dollars to the American economy. The most significant production of America was corn and sorghum. The second-largest producer of agricultural products in the world is China, which sold 230.38 million tons of agricultural products to the market last year. This populous Asian country, which is America's main competitor in the economic field, has been one of the largest producers of agricultural products throughout history. Of course, it is the reason for providing food for more than 1 billion people in this country.

Studies conducted by the World Food and Agriculture Organization show that only a few countries have produced agricultural and food products successfully. A fundamental reason for this success can be seen in the need for land to produce agricultural products. As a result, countries with a large area and enough land for agricultural work can be prosperous in the agricultural industry and provide products for the global market. In this report, the World Food and Agriculture Organization has introduced four countries as the world's largest producers of food products: China, India, America, and Brazil.

These countries are pioneers in producing agricultural products and have a special place in exporting food products globally. Among these countries, America can be introduced as the most influential country in the production of food products in the world. This country is the world's largest producer of food products and the largest exporter. But China and India, among the largest producers, cannot export a large share of these products to the world market due to their large populations and high domestic consumption. Another large country that was active in the production of food products is Brazil. This country is active in producing sugarcane, soybeans, and beef; in these sectors, it has gained a special place in the world. We used to introduce the food industry, but the fact is that this country is not among the largest producers of agricultural products in the world, and the reason for this is the unfavorable weather conditions of this country, which has destroyed the possibility of producing agricultural products. A large part of the land in Russia is uncultivable, and it is impossible to produce agricultural products on them.

China is one of the largest producers of agricultural products and food products in the world. But due to the large population and high consumption needs, China is forced to import food products to meet domestic needs. Even though a large part of China's land is in the mountainous region, it can be used for rainfed agriculture because of the soil quality. China can supply many domestic needs by producing on domestic farms, and the quality of agricultural products is impressive among the world's products.

Another point that has made China successful in the food production and agriculture industry is the high population of this country and the large workforce active in agriculture. According to the official reports of the Chinese government, 315 million people work in the food and agriculture industry, which is more than the entire population of the United States. This Asian country produces highly valued food products, and their demand is very high. Among these agricultural products, rice and wheat can be mentioned. China is one of the largest producers of vegetables, such as potatoes, lettuce, onions, cabbage, green beans, broccoli, eggplant, spinach, carrots, cucumbers, tomatoes, and pumpkins, and in the production of fruits such as pears, grapes, apples, peaches, and plums. And watermelon is very active. This country has considerable activity in the livestock and livestock products sector. Statistics show that China is one of the largest producers of sheep's milk, chicken, beef, goat, and fish and consumes many eggs and honey. This country is one of the largest producers of peanuts in the world.

The agricultural industry in America is one of the most efficient industries, and the productivity of this country is very high. This country has a smaller labor force than India and China, but the production volume of American agricultural products is almost equal to that of China. The World Food and Agriculture Organization has announced that the reason for the high production volume of agricultural products in the United States is the country's use of new production technologies and methods, which have increased production efficiency.

The important thing about the agricultural industry in America is that farms are found in different parts of the country, and not just one region is engaged in producing agricultural products. But some states play a more significant role in producing essential agricultural products than others, including California, Iowa, Texas, Nebraska, Minnesota, Illinois, Kansas, North Carolina, Wisconsin, and Indiana. The main agricultural products produced in this country are corn, soybeans, wheat, cotton, tobacco, rice, sorghum, and barley. Still, the chicken is also very active in producing various fruits and vegetables, including grapes, oranges and apples, lettuce and cabbage, soybeans, and turkey meat.

American food companies also play an important role in exports, and a large part of this country's products is exported to the markets of Canada, Mexico, China, Japan, and Germany. The main export products of the United States are soybeans, wheat, and corn. Other export products of this country include almonds, pistachios, cotton, potatoes, and other poultry meat products. According to the statistics provided by the U.S. Department of Agriculture, there are 2.2 million large farms in America, and more than 93% of these farms use the latest technologies and science in the world to produce what they pay. America has been introduced as the country where the agricultural industry has the highest efficiency.

India is another country that is famous in the world for producing agricultural products and food. The share of agriculture in this country's GDP is 15.4%. This country has vast farms, but India's agricultural land area is less than that of China, America, and Brazil. India plays a more active role in exporting agricultural products than China because many people living in India are so poor that they cannot buy food and agricultural products. For this reason, the products produced in this country enter the world markets.

On the other hand, the population growth rate in this Asian country is very high, and the absolute poverty in many parts of India has fueled the rapid population growth in this country. The United Nations reported India's economic performance over the last two decades. In this report, they mentioned, "The economic growth rate in this country is high, and the opportunities for economic growth in this country are very high, but our concern is that the population growth rate is too fast. Economic growth should be higher, and many opportunities will be lost."

In 2019, the World Bank announced the population of India as 1.37 billion people. Still, due to this country's high population growth rate, it is expected that in the not-too-distant future, this country will become the most populous country in the world and surpass China in this regard.

India's leading agricultural product is rice, worth more than 100 billion dollars annually. This country's other valuable agricultural products include buffalo, cow, wheat, cotton, mango, and all kinds of fresh vegetables, including potatoes, tomatoes, and onions. This country is essential in producing bananas, sugarcane, soybeans, and peas. This country's agricultural land is under rice cultivation, followed by wheat, oilseeds, sugarcane, tea, and cotton. Figure 1.1 presents the share of production of each continent.

According to the information we obtained from the Ministry of Agricultural Jihad in Iran, the agricultural sector is responsible for providing food needs by relying on national production and optimal and efficient use of production resources and protection of renewable natural resources and increasing farmers' income. With about 6.6% of GDP, 17.7% of employment, and 5.9% of non-oil exports, this sector provides about 80% of food and 80%–90% of raw materials needed by the country's industries [15].

Iran has a potential of 37 million hectares of arable land, 120 million livestock units, 84.8 million hectares of pastures, 14.3 million hectares of forests and rich genetic resources, about 242 billion cubic meters of rainfall, 104 billion cubic meters of surface water resources, accessible underground, 2,700 kilometers of maritime borders in the north and south, and 4.2 million beneficiaries having a significant scientific, expert force in the agriculture and natural resources sector. Finally, climate diversity can be achieved with integration and coordination. It is necessary to promote the appropriate productivity of the mentioned resources to be considered one of the region's patterns of comprehensive agricultural development [16].

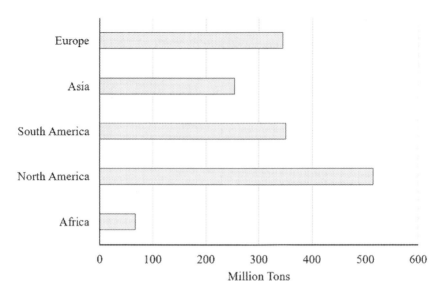

FIGURE 1.1 The share of agricultural production.

Agricultural and food products are classified into crops, horticulture, live-stock, and fishery. As shown in Figure 1.2, although the trend of crop production was upward during 2009–2021, it experienced fluctuations during 2015–2021. During the years under review, the country's most significant agricultural production was allocated to wheat; after that, fodder corn had the largest share among agricultural products.

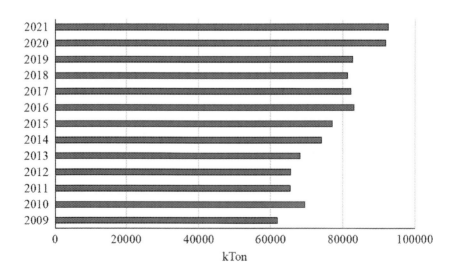

FIGURE 1.2 The agricultural production of Iran during 2009–2021.

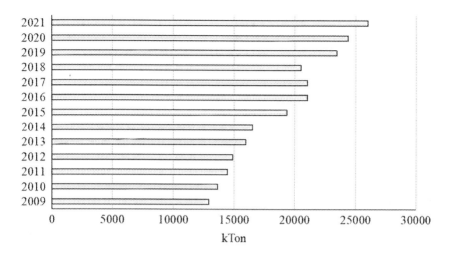

FIGURE 1.3 The horticulture production in Iran during 2009–2021.

As shown in Figure 1.3, the trend of garden production in Iran has been increasing. In addition to family consumption, garden products are essential in increasing the country's exports and foreign exchange earnings and providing the necessary resources to improve economic access to food. Citrus, stone, and apples are essential garden products.

In Iran, protein products play an essential role in food security. As shown in Figure 1.4, the trend of livestock production during the years 2009–2021 has been increasing. Livestock and poultry products can be classified into red meat, chicken, eggs, milk, and honey.

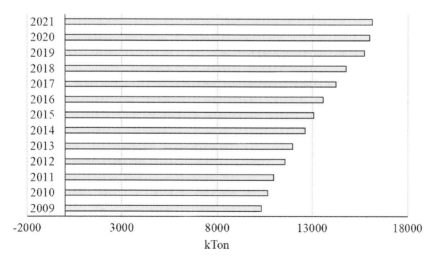

FIGURE 1.4 The livestock production in Iran during 2009–2021.

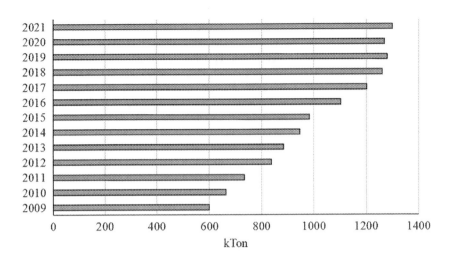

FIGURE 1.5 The fishery production of Iran during 2009–2021.

In Iran, fishery products can be classified into two categories: fishing and aquaculture. Figure 1.5 shows that fisheries and aquaculture production trends increased during 2009–2021. Fisheries and aquaculture productions have played an essential role in health and food security and have attracted the attention of countries due to fewer emissions of greenhouse gases. Using open and international waters will reduce some side costs of its production.

According to the studies conducted, it can be said that agricultural development is of great importance, as it provides food security in the domestic society, develops exports, causes entrepreneurship and income generation in rural communities, and preserves natural resources and the environment. It can be done and also plays an essential role in eliminating poverty and diminishing class differences. By investing in the agricultural sector, in addition to personal profit, we find a significant contribution to the prosperity of our country's economy. Although the drought and the lack of water resources have caused the agricultural sector to face many problems in recent years, the use of modern irrigation methods and the development of greenhouse cultivation, and extensive studies in the field of water and soil have primarily solved the problems and strengthened. The development of the agricultural sector has led.

1.3 CONCERNS, POSSIBLE SOLUTIONS, AND OPPORTUNITIES

The agricultural industry in developed countries such as North American and European countries is managed intelligently in almost all sectors. This industry also uses innovative capabilities in developing countries in its various sectors. Using new methods and technologies can reduce resource consumption in agricultural lands. The most important thing is the drastic reduction of water consumption, which is very important for desert countries. Iran has much more limited water resources than European and North American countries. However, irrigation is still done traditionally in many areas of Iran, which causes a lot of wastage of this valuable resource. In recent years, drip irrigation has grown relatively large in the country. Still, along with drip irrigation, other agricultural industry technologies can further reduce resource consumption, especially water.

Another big industry problem in Iran is how to sell agricultural products. Agricultural products are sold in Iran in traditional markets, mainly controlled by brokers and not closely monitored. This method has caused the agricultural product to be bought and sold from the land to the final customer by several brokers. The quality reduction caused by a long supply chain increases the final price. Using new technologies and platforms based on the Internet and mobile phones can cause a massive transformation in the agricultural industry. These developments can cause an economic revolution in Iran because Iran can grow all kinds of agricultural products and export them, which can cause economic growth at the macro level. This article investigates the challenges and appropriate solutions have been presented to solve the related challenges.

In developing countries, one of the fundamental problems in the agriculture and livestock industry is buying and selling products. Farmers make the most effort in this industry, and a tremendous amount of risk is related to farmers. However, they get the most negligible financial benefit because the management of buying and selling agricultural products is controlled by a vast network of intermediaries in this field. The suggested solution to deal with this problem is direct sales or with a minimum number of intermediaries. Today, the Internet and smartphones connected to the Internet have allowed all citizens to buy their desired products online. This technology has led to the creation of various online stores. Farmers can sell their products at a fairer price and make more profit by selling their products in the stores they create. It will remove all the intermediaries in this path. But setting up an online store with a high capacity to sell agricultural products is not easy; even so, farmers will need logistics facilities that they usually do not have. Therefore, companies have been formed to sell agricultural products without intermediaries. These companies themselves are, in a way, sales intermediaries. Still, by meeting the needs of farmers and increasing their profit margins, they have been able to help eliminate additional intermediaries in this industry. Therefore, the number of these stores and the development of their basic infrastructure can control the price, eliminate intermediaries, facilitate sales and increase farmers' income.

Today, new technologies are used in all industries. In the advanced countries of North America and Europe, modern agricultural methods have reduced the consumption of primary resources such as water, fertilizers, poisons, pesticides, etc. In Iran, traditional methods are still used in most regions. One of the things that have been done in recent years is drip irrigation. However, using this method is only effective in reducing water consumption. Still, by using modern agricultural methods, in addition to further reducing water consumption, other resources can also be optimized. Some of the new agricultural methods are listed below:

A. Internet of Things: The Internet of Things has created a significant transformation in most different industries. The agricultural industry will not benefit from this technology either. Using a network of sensors on the ground and collecting sensor data provides the farmer with information that can be used to make more effective decisions. Also, using the information extracted from these sensors, many activities such as irrigation, spraying, fertilization, and pesticides can be automatically converted. In addition to significantly reducing the farmer's physical activity, this work will also optimize the consumption of essential resources.

B. Aerial photography: One of the new services offered to farmers worldwide is providing advice and analysis based on aerial photography. This type of imaging is generally done using satellite imaging or using drones. These images indicate which part of the land needs more irrigation, fertilization, spraying, or pesticides. Carrying out activities based on satellite images will reduce the consumption of primary resources on the level of agricultural land.

C. Smart greenhouses: These greenhouses are made semi-automatically and fully automatically. Usually, the production capacity in these greenhouses is several times that of traditional agricultural lands. All the operations related to the use of resources in greenhouses are done automatically and using this work, the use of resources in these greenhouses is entirely optimal.

D. Modern farming methods: Today, traditional farming methods are becoming obsolete worldwide. New methods such as vertical farms, hydroponic, and aeroponic cultivation reduce space and water requirements by 95% and increase outcomes up to 300 times.

On the other hand, increasing the scientific level of farmers is also felt to deal with challenges and obstacles. Social networks and consulting platforms based on the Internet and smartphones can be used to increase farmers' awareness of new technologies. But one of the practical ways can be implementing pilot projects by start-ups to encourage farmers to use new technologies by presenting documented reports of increasing product productivity.

There are many challenges in the agriculture industry, including:

1. Using traditional agriculture and wasting inputs and resources
2. Expensive modern equipment
3. High effectiveness of agricultural and livestock products against environmental conditions
4. Integrated and efficient management complexity
5. Multiple sales intermediaries
6. Low productivity due to uneducated farmers
7. Low productivity and high wastage due to a lack of sufficient information about market needs

8. Changing consumer preferences to use organic and fresh ingredients
9. Natural resources limitation and destructive environmental effects of extensive agricultural activity
10. Labor shortage and increased personnel costs

Table 1.1 presents some possible solutions to solve the mentioned problems.

1.4 DIGITAL FARMING

This section discusses the analysis and interpretation of the three main key words frequently used in this book: precision farming, smart farming, and digital farming. As shown in Figure 1.6, digital farming includes precision farming and smart farming. The process starts with measurement and ends with the application. In other words, digital farming creates value from data [17]. It is "the application of both sections, internal and external networking of the farm and use of web-based data platforms together with Big Data analyses" [18]. Precision farming can be defined as "a farming management technology to observes, measures, and analyzes the needs of fields and crops based on Big Data, analytics capabilities, and robotics/aerial imagery, sensors, and weather forecasts" [19]. It is "a modern farming management concept using digital techniques to monitor and optimize agricultural production processes" [19]. Smart farming is the application of information and data extracted from precision farming. It focuses on how the data can be used in smart farm operations [20].

1.5 BUSINESSES

Digitalization trends, such as the advancement of web accessibility, have produced new tools for enhancing creative techniques of business advertising, and organizations are creating websites to take advantage of the Internet's commercial potential [21]. The new era of the digital economy has had an impact on marketing as well. Digital marketing was first developed in the 1990s and early 2000s, and it firmly emphasizes using the Internet and other digital technology to advertise goods and services [22]. Businesses based on digital technologies are necessities of the rural and agricultural production system and have different dimensions. Knowledge is necessary for every wise move and moral decision. Knowing the market is a systematic effort to collect, record, and record information related to all the constituent parts of the market system. Business promotion requires marketing. Marketing means penetrating the market and introducing and making the organization's products and services known. Marketing is keeping the customer and creating an environment to bring him back to himself by constantly communicating with people, providing appropriate services, and being aware of the competitors' movements [23]. According to what was said, marketing is a dynamic and multi-dimensional process that requires a systemic approach.

Success in a competitive environment requires increasing the quality of products and management processes while reducing costs. However, studies show that

TABLE 1.1

Concerns, Possible Solutions, and Opportunities

Challenge No.	Possible Solution	Parties and Customers	Position in the Activity Chain	Examples
1	Water and nutrients continuous measurement system of a plant using in-ground sensor	C_2	P_1	• Urban farming systems: vertical cultivation, hydroponics, aeroponics
	Reducing water consumption using new planting systems	C_1		• Smart irrigation and fertilization system
	Optimizing water and fertilizer consumption using robotics and IoT	C_2		
2	Online platforms for renting tools and equipment	C_2	P_4	• Online shops and communication systems
	Online platforms for buying and selling used and second-hand equipment			• A platform for selling household products
3	Continuous monitoring systems	C_1	P_1	Urban farming systems: vertical cultivation, hydroponics, aeroponics
	Smart weather forecast and quick notification	C_2	P_2	Information and decision services
4	Facilitate and plan activities (from land preparation to product sales), resource allocation, information on product status, weather, market, and integrated management	C_2	P_1	Aerial photography
	The automatic recording system of product status and ongoing activities	C_2		
	Forecasting of harvest volume using modern systems and data analysis	C_8	P_2	• Information and decision services • Management dashboards
5	Comparison, consultation, and purchase of equipment and raw materials using online platforms	C_5	P_2	Platform for wholesale
	Direct wholesale or retail using online platforms	C_6		
	Facilitating marketing activities and informing sales exhibitions using online platforms	C_2	P_3	Promotion of skill training courses
	Connecting different parties through a specialized network	C_7	P_4	• Online shops and communication systems • A platform for selling household products

No	Description	C	P	Details
6	Teaching agricultural and livestock skills using online and offline platforms	C_2	P_3	Promotion of skill training courses
	Online and offline training-promotional courses about modern agricultural and animal husbandry equipment utilization			
	Encouraging and promoting synergy of collective learning and sharing of successful experiences and creation of a specialized network			
7	Water and nutrients continuous measurement system of the plant using in-ground sensor	C_1, C_2	P_1	• Urban farming systems: vertical cultivation, hydroponics, aeroponics
	Plant cultivation in controlled conditions	C_1, C_2		• Aerial photography
	Identification and analysis of plant needs using aerial images	C_1, C_2		
8	Plant cultivation in controlled conditions	C_1	P_1	Urban farming systems: vertical cultivation, hydroponics, aeroponics
	Establishing cultivation fields (greenhouse or non-greenhouse) near stores and distribution centers to maintain product freshness	C_1, C_2		
	Promoting education and providing required equipment to plant at home or the place of consumption	C_4	P_3	Promotion of skill training courses
	Platform for selling	C_4	P_4	• Online shops and communication systems
				A platform for selling household products
9	Increasing productivity by dimension changing (horizontal to vertical) and using urban environments	C_1	P_1	• Urban farming systems: vertical cultivation, hydroponics, aeroponics
	Reducing the use of water resources and using modern systems	C_1		• Cell culture and laboratory products
	Promoting education and providing required equipment to plant at home or the place of consumption	C_1	P_3	Promotion of skill training courses
	Production of livestock products with cell culture	C_3		

(*Continued*)

TABLE 1.1 (Continued)
Concerns, Possible Solutions, and Opportunities

Challenge No.	Possible Solution	Parties and Customers	Position in the Activity Chain	Examples
10	Making it possible to search, compare, and temporarily hire seasonal farm workers online	C_2	P_1	• Urban farming systems: vertical cultivation, hydroponics, aeroponics
	Reducing workforce requirements using robotics and machine vision technologies	C_2		• Smart irrigation and fertilization system
				• Automation, robotics, and unmanned tractors
	Converting traditional tractors to self-driving tractors	C_2	P_2	A platform for hiring temporary workers
	Automated agricultural activities in modern farming systems and closed environments	C_1		

Parties and customers
C_1: Greenhouse owners, urban, and household farmers
C_2: Farmers and ranchers with various income ranges
C_3: Industrial and laboratory producers of agricultural and livestock products
C_4: Household farmers
C_5: Suppliers
C_6: Consumers
C_7: Stores and retailers
C_8: Government and policy-making institutions

Position in the activity chain
P_1: New farming methods and precision farming
P_2: Facilitate management activities
P_3: Knowledge and network improvement
P_4: Online services

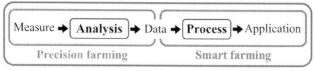

FIGURE 1.6 Illustration of digital farming and its subsections.

although companies can benefit from information technology, they need special conditions to benefit from these benefits and be successful in using this technology and are affected by several factors [24]:

1. *Social and economic readiness:* refers to the degree and extent to which a community is prepared to participate in developing the digital economy. The readiness to accept digital technologies is different among different societies. All society's social and economic infrastructures should be in a position to support the acceptance of digital technology.
2. *Infrastructures*: Infrastructures are the central pillar in the marketing of agricultural products, which include:
   A. Information and communication technology infrastructure
   B. Transportation facilities or local distribution channels
   C. An accessible and reliable source of electricity, especially in rural areas.
3. *Knowledge and literacy level.*
4. *Fair and equal access*: fair and equal access to infrastructure, correct use of information and communication technology, and fair allocation of information and communication technology resources are necessary to achieve positive and desirable results.
5. *Acceptability and compatibility:* communication tools and tools such as mobile phones, messengers, personal computers, and Internet sites must be culturally acceptable and compatible with society.

You will see a big revolution in the agriculture industry in the next few years. Robots, sensors, irrigation, and fertilization systems are being tested worldwide. With the professional entry of this technological tool into the modern world, the agricultural industry's future will undergo considerable change. With the advancement of the Internet of Things (IoT), many agricultural affairs can be controlled only with a mobile phone. Using the IoTs, you can easily connect to the central system with your mobile phone and control the robot, sensor, camera, irrigation system, fertilization, or anything else you have. Among the essential businesses related to digital agriculture, the following can be mentioned:

- Application of global positioning system (GPS) in agriculture
- Application of geographic information system
- Application of intelligent irrigation and fertilization system

- Application of robotic agriculture
- Application of smart sensors in agriculture

1.5.1 APPLICATION OF GLOBAL POSITIONING SYSTEM (GPS) IN AGRICULTURE

GPS in agriculture has made it very easy to locate and map land. As you know, nowadays, there are GPS receivers in all smartphones. It has caused the price of these receivers to drop drastically and their use to increase. Using these receivers, you can easily communicate with the central system of your agricultural land and do things like irrigation and fertilization.

1.5.2 APPLICATION OF GEOGRAPHIC INFORMATION SYSTEM

What is the geographic information system, and what is its use in agriculture? The answer to this question is straightforward. This intelligent system is used in agricultural areas and maps agricultural land digitally. In this mapping, information such as land topography and soil survey makes it easier to analyze the soil and production. In addition, it makes it easier to identify soil sources. It will help farmers to find the right crop to plant on their farmland and provide the right items for planting.

1.5.3 APPLICATION OF INTELLIGENT IRRIGATION AND FERTILIZATION SYSTEM

Seeds need regular care, watering, and fertilizing after planting until harvest. Also, the seeds need moderate weather and sunlight to grow. The intelligent irrigation and fertilization system with sensors can easily consider all these points. The water, air, light, and weather conditions introduce the right amount of water and fertilizer into the soil for plant growth.

This system will minimize the spoilage of fruits in the cold and hot seasons and increase the amount of food on the soil surface. These sensors are connected to a central system. You can also connect to this system with your mobile phone and change the settings or start watering and fertilizing. The innovative irrigation system has been implemented in some countries and has had good results. Currently, this system is used in some smart villages. Using this system, you can manage your agricultural land even if you travel to another country.

1.5.4 APPLICATION OF ROBOTIC AGRICULTURE

The arrival of gardener robots in smart agricultural fields can be considered a revolution in this industry. The entry of robots into the agricultural industry can increase the speed of planting and harvesting crops and cause them to grow better. Currently, robots are built in the form of cars that plant seeds and take care of them daily until it is time to harvest. These robots examine all aspects of the soil and choose the best spot for planting seeds. Another robot made in this field is in the form of a drone and is used for spraying. This intelligent robot spreads fertilizer and pesticide accurately and in a tiny form on plants and flowers. Recently, a drone was unveiled that can intelligently cut the poison down to the size of nano-sized materials and spray it on plants.

1.5.5 APPLICATION OF SMART SENSORS IN AGRICULTURE

It is interesting to know that in modern agriculture, sensors are placed around the earth where image recognition technologies are used. These sensors allow farmers to know the status of their crops anywhere in the world. These sensors send updated information to the farmers in real time to make the necessary changes to their crops. Imagine having an app that tells you when plants in your yard need water and other nutrients. In this case, many of the problems you face in growing plants will disappear. IoT sensors placed inside fields do the same for farmers. Of course, the Internet of Things does this on a larger scale, which causes the production of more food and less water consumption.

1.6 LABORATORIES

Due to the world's population growth and the increase in demand for agricultural products, today, the world community is facing a severe food crisis. Finding new sources for producing agricultural products is one of the practical solutions to this crisis. Smart agriculture is considered one of the best ways to deal with the global food crisis and a new and cost-effective method in agriculture. Among the advances made in agriculture is establishing centers and laboratories related to precision and digital agriculture worldwide. Table 1.2 presents these centers' purpose and title examples.

1.7 START-UPS

For the past few decades, start-ups have tried to improve the agriculture industry by identifying problems and providing valuable and practical solutions. In this section, the presented global start-ups are classified into four sections: digital farming, management, online services, and knowledge and network improvement.

1.7.1 START-UPS IN DIGITAL FARMING

These start-ups are more focused on: (1) the effects of climate on the process of agricultural activity, (2) matching market needs with suitable products for cultivation, (3) reducing wastage of products and inputs, (4) labor supply management, (5) resource limitation, and (6) unified and optimized management. Table 1.3 presents the details of start-ups in this section.

1.7.2 START-UPS IN FIELD MANAGEMENT ACTIVITIES

These start-ups are more focused on: (1) the effects of climate on agricultural outputs, (2) labor management, (3) eliminating/reducing the influence of intermediaries

TABLE 1.2
Some Research Laboratories Focused on Smart Agriculture

Name	Country	Mission	Websites
The University of Sydney Institute of Agriculture (PA-Lab)	Australia	To lead precision agriculture science and manage agricultural industries for sustainable production	https://precision-agriculture.sydney.edu.au/
Robotics and Internet-of-Things (IoT) for Smart Precision Agriculture and Forestry	Portugal	To optimize the intelligent precision agriculture and forestry, profitability, and automation	https://www.inesctec.pt/en/laboratories/laboratory-of-robotics-and-iot-for-smart-precision-agriculture-and-forestry#contacts
Laboratory of vehicle robotics	Japan	ICT, robots for innovation in 21st-century agriculture	https://vebots.bpe.agr.hokudai.ac.jp/en/
National Center for International Collaboration Research on Precision Agricultural Aviation Pesticides Spraying Technology	China	Precision agriculture for sustainable production	https://english.scau.edu.cn/432/list.htm
National & Local Joint Engineering Research Center for Precision Processing of Livestock and Poultry Products and Safety Control Technology	China	To provide sustainable livestock and poultry products and employ Safety Control Technology	https://english.scau.edu.cn/432/list.htm
Precision Agriculture Laboratory	USA	To reduce waste, increase profits, and maintain the quality of the environment	https://abe.ufl.edu/precag/
Guo Precision Ag Lab	USA	To provide a sustainable production system in agriculture	https://www.depts.ttu.edu/pss/PrecisionAg
Digital Laboratories and agriculture	South Africa	To monitor the effectiveness of agricultural productions	https://agribook.co.za/services-and-technologies/laboratories-and-agriculture/
Electromagnetics Precision agriculture	United Kingdom	To overcome the imaging challenges for agri-technology	https://www.npl.co.uk/electromagnetics/precision-agriculture
Precision Agriculture Research Solutions	United Kingdom	To identify adulteration and monitor crop maturity	https://analytik.co.uk/precision-agriculture-research-solutions/
Precision farming source	Ukraine	Agrochemical soil analysis for companies	https://agrotest.com/en/services/precision-agriculture/

TABLE 1.3
Examples of Agricultural Start-ups in the Field of Smart Farming

Name	Year	Country	Product	Income Model	Key Tech.	Ref.
AeroFarms	2004	USA	Urban aeroponic cultivation system and online sales of products	Direct sale	Sensors, automation, online platform	[25]
AirWood	2014	India	Comprehensive agricultural service system from seed selection to harvesting	Service	Aerial image processing, software, and online platform	[26]
Alesca Life	2016	China	Urban hydroponic cultivation system and online sales of products	Direct sale	Sensors, automation, online platform, and software	[27]
Arable	2013	USA	Plant status monitoring system	Subscription fee (application), direct sale (instruments)	Data mining, AI, software	[28]
Bitwater	2015	USA	Automation system and control of environmental conditions of insect cultivation	Direct sale	Sensors, automation, artificial intelligence	[29]
Blue River Technology	2011	USA	Automatic precise weed spraying system	Direct sale	Machine vision, data mining, AI	[30]
BrightFarms	2011	USA	Greenhouse cultivation system near big stores	Direct sale	Greenhous system, sensors, automation, and robotics	[25]
CropX	2013	Israel	Water and fertilizer monitoring system	Subscription fee (application), direct sale (instruments)	Sensors, IoT, software, online platform	[31]

(Continued)

TABLE 1.3 (Continued)
Examples of Agricultural Start-ups in the Field of Smart Farming

Name	Year	Country	Product	Income Model	Key Tech.	Ref.
Ceres Imaging	2014	USA	Agricultural land imaging system with airplane and its analysis	Service	Aerial image processing	[32]
Clearpath Robotics	2009	Canada	Robotic technology, apple harvester	Direct sale, service	Automation and robotics, sensor and machine vision, unmanned vehicle	[33]
Droplet	2014	USA	Smart watering robot	Direct sale	IoT, robotic	[34]
Digital Spring	2013	USA	Moisture monitoring system for pots and garden plants	Direct sale	Sensor, IoT, software	[35]
eFarmer B.v.	2016	India	The system of converting traditional tractors into self-driving tractors	Subscription fee (application), direct sale (instruments)	GPS, unmanned vehicle	[36]
FreshBox Farms	2015	USA	Providing agricultural products using vertical and urban cultivation systems	Direct sale	Sensors, automation, online platform	[37]
Freight Farms	2010	USA	Hydroponic and vertical cultivation	Direct sale	Sensors, automation, online platform	[38]
GardenPlug	2016	USA	Soil-less cultivation pots	Direct sale	Sensors	[39]
HoneyComb Corporation	2012	USA	UAV control application and aerial image analysis	Direct sale	Aerial image processing, software, and online platform	[40]
Harvest croo	2013	USA	Strawberry harvester	Direct sale, Service	Sensor, automation, data mining, AI	[41]
Impossible Foods	2011	USA	Burgers made with lab meat	Direct sale	Biotechnology and cell culture	[42]

Company	Year	Country	Description	Business model	Technology	Ref.
Indigo Agriculture	2014	USA	Seeds with microbes that protect the plant against environmental conditions	Direct sale	Biotechnology	[43]
Orbital Insight	2013	USA	Forecasting system of agricultural products by analyzing aerial images	Subscription fee	Aerial image processing, software	[44]
Perfect Day Foods	2014	Ireland	Artificial milk	Direct sale	Biotechnology and cell culture	[45]
Raicho	2012	USA	Intelligent garden irrigation control system	Direct sale	Software, online platform, automation	[46]
Skycision	2015	USA	UAV control application and aerial image analysis	Subscription fee	Aerial image processing, software	[47]
Smart Ag	2016	USA	The system of converting traditional tractors into self-driving tractors	Subscription fee (application), Direct sale (instruments)	IoT, unmanned vehicles, machine vision	[48]
TerrAvion	2013	USA	Aerial imaging system and its analysis	Subscription fee	Aerial image processing	[49]
Tevatronic	2012	Israel	3D soil monitoring system and automatic irrigation	Direct sale	Sensor, IoT, automation	[50]

on the final price of products, and (4) division of duties and management of activities. Table 1.4 presents the details of start-ups in this section.

1.7.3 START-UPS IN ONLINE SERVICES

These start-ups are more focused on: (1) increasing the quality of products offered to consumers, (2) eliminating/reducing the influence of intermediaries on the final price of products, and (3) the price of equipment and cost-effectiveness of using them for farmers. Table 1.5 presents the details of start-ups in this section.

1.7.4 START-UPS IN FARMERS' KNOWLEDGE AND NETWORK IMPROVEMENTS

These start-ups are more focused on: (1) Creating communication between farmers/herders/related people/ ..., (2) knowledge improvement of farmers and responsible people, and (3) information about new technologies. Table 1.6 presents the details of start-ups in this section.

1.8 NEW IDEAS ON DIGITAL FARMING

Today's agriculture is very advanced and uses technologies such as robots, temperature and humidity sensors, aerial images, and GPS technology. These advanced devices, precision agriculture, and robotic systems enable businesses to be more profitable, efficient, safer, and environmentally friendly.

The latest techniques and superior technologies in agriculture are:

- GIS software and agriculture with GPS
- Satellite imaging
- Drone
- Aerial imagery
- Agricultural software and online data
- Data set integration

From the effects of technology on agriculture, it can be said that technological innovations have always given a better shape to agriculture, from ancient times when plows were used, to today's precise and advanced equipment and global positioning systems (GPS). It has provided humanity with new methods to achieve greater efficiency and tremendous food growth.

Among the advantages of new technologies are the following:

- Modern machines are used to control the efforts of farmers
- Technology in agriculture reduces the time spent on different activities
- Machines are used to plant seeds
- Irrigation technology will be faster and easier
- Chemical pest control is done better
- Soil fertility improves

TABLE 1.4
Examples of Agricultural Start-ups in the Field of Management

Name	Year	Country	Product	Income Model	Key Tech.	Ref.
Agrimap	2006	New Zealand	Farm management software	Subscription fee	Software, online platform	[51]
Bovcontrol	2012	Brazil	Online platform for collecting and analyzing the information of livestock farmers	Subscription fee	Online platform	[52]
Cheruvu	2014	India	Data mining and artificial intelligence platform for agricultural activities	Direct sale	AI, data mining, online platform, software	[53]
FarmLogs	2012	USA	A platform to manage and plan agricultural activities	Subscription fee	Software, online platform	[54]
Farmnote	2006	Kenya	A platform to provide the required information to farmers in disadvantaged areas	Direct sale	AI, data mining, online platform, software	[55]
Ganaz	2017	USA	A platform for searching and hiring farm workers and communicating with them	Subscription fee	Software, online platform	[56]
Prospera Technologies	2014	Israel	The platform for forecasting production and market conditions	Subscription fee	Sensor, data mining, online platform	[57]
Stellapps	2011	India	Monitoring system of dairy supply chain	Subscription fee, direct sale	Online platform	[58]
Tracker.com	2012	Germany	A platform to manage and plan economic and managerial activities	Subscription fee, freemium	Software, online platform	[59]
Tambero	2014	Argentina	A platform to manage and plan agricultural activities	Subscription fee	Software, online platform	[60]

TABLE 1.5
Examples of Agricultural Start-ups in the Field of Online Sales

Name	Year	Country	Product	Income Model	Key Tech.	Ref.
Aggrigator	2015	USA	A platform for buying and selling fresh local products	Intermediation	Software, online platform	[61]
Agriconomie	2014	France	Online platform for purchasing consumables needed by farmers	Intermediation	Software, online platform	[61]
AgroStar	2008	India	Online platform for purchasing consumables needed by farmers	Intermediation	Software, online platform	[61]
BumperCrop	2012	USA	A platform for buying and selling small farm products	Intermediation	Software, online platform	[62]
EM3 AgriServices	2013	India	Advanced agricultural equipment rental platform	Subscription fee	Software, online platform	[63]
La Ruche Qui Dit Uui	2010	France	Online platform for direct sale of agricultural products	Intermediation	Software, online platform	[61]
Meicai	2014	China	A platform for restaurants to buy fresh ingredients from farmers	Intermediation	Software, online platform	[64]
MissFresh	2014	China	Online retail of agricultural and livestock products	Intermediation	Software, online platform	[65]
Machinio	2013	USA	Second-hand agricultural machinery buying and selling search engine	Intermediation	Software, online platform	[66]
ProduceRun	2014	USA	An online platform for farmers to carry out marketing activities	Intermediation, direct sale	Software, online platform	[67]

TABLE 1.6

Examples of Agricultural Start-ups in the Field of Networking and Knowledge Improvement

Name	Year	Country	Product	Income Model	Key Tech.	Ref.
Farmers Business Network	2014	USA	Specialized and comprehensive agricultural network	Subscription fee	Software, online platform	[68]
Grow the Planet	2011	Italy	Training and empowerment platform	Direct sale	Software, online platform	[69]
LivestockCity	2014	USA	Online platform for information and virtual services for livestock farmers	Subscription fee	Software, online platform	[70]
Livestock Connect	2013	Australia	A platform for connecting different livestock sector activists	Subscription fee	Software, online platform	[71]

- The use of water and chemical fertilizers decreases and reduces the final price
- Chemicals are used less, and waste materials fall into the seas less

Among the disadvantages of new technologies in agriculture, the following can be mentioned:

- Excessive use of chemicals reduces soil fertility
- Farmers cannot handle machines and equipment properly due to a lack of practical knowledge
- Sometimes excessive use of machinery leads to environmental damage
- Since in some parts of the world, farmers cannot use modern machinery

The advent of electronics has tremendously impacted modern agriculture and equipped farms and greenhouses with various sensors, devices, and analyses. Some many tools and machines support this sector, but to understand this issue, they can be classified into different categories:

- *Electronic equipment:* There are various sensor combinations such as a pyranometer (measurement of sunlight) or soil tester that help the farmer understand the deficiency or excess of the element, and robotic equipment to assist with manual tasks.
- *Machinery:* Modern machines with multiple capabilities can replace busy jobs and make them more accessible, cheaper, and safer.
- *Software and data science:* Various plant physiology, soil health analysis, and event management software such as Insight Manager are used in this field, which can help the farmer stay up to date and plan the farming strategy along with modern data science. It ultimately helps the growth of agriculture.
- *Plant biotechnology:* This section is about rejuvenating the crop and ensuring you have a strong foundation for the plant.

As your farming business grows, you will need more investment, labor, and resources. These limiting factors are managed with modern agricultural methods to ensure profit. Resources such as fertilizers, pesticides, irrigation equipment, and machinery should be spent on resources, etc.; but the required costs are minimized by using modern methods. For example, in the middle of growth, calcium nitrate decreases in the soil of agricultural land and reduces the yield; but there are sensors in your soil that warn of this decrease. If there is no such equipment, the production performance will decrease, and as a result, the sales efficiency and income from agriculture will decrease.

It is interesting to know that the impact of modern agriculture is more on medium and small farmers. It is because more control over the farming space can make them more productive than industrial-scale farmers who have already implemented modern technology. By reducing the dependence on chemical agents, the food produced becomes fresher and healthier. The high health of these products and easy access increases the demand. In addition, the need for labor and irrigation is significantly reduced, and the yield of agricultural land is surprisingly high.

Modern agriculture has a significant effect on increasing production and, thus, increasing profits. But apart from these, the main advantage for everyone is "sustainability." In this method, none of the methods are destructive or abrasive. Therefore, fertile soil is protected, water consumption is saved, natural resources are used efficiently, and healthy and fresh products are obtained. Modern agriculture provides all plants' needs, and resources are not wasted. As the Internet of Things grows, manual intervention will be minimized, and ultimately profits will be high.

Types of new ideas in today's modern agriculture include the following.

1.8.1 INDOOR VERTICAL FARMING

Indoor vertical farming reduces the distance traveled in the supply chain and increases the yield. Indoor vertical farming is a method of growing crops in a closed, controlled, and stacked environment. Growing racks installed vertically can significantly reduce the space required. This type of plant cultivation is usually done in a limited urban space. These farms are also unique in another way: the lack of soil. Therefore, they are helpful for plants that do not need soil for plant growth. Vegetables are usually grown in a bowl of nutrient-rich water, or the plant roots may be systematically sprayed with water and nutrients. Artificial growth lights are also used instead of natural sunlight.

The benefits of vertical farming include the following:

- sustainable urban growth;
- maximum product performance;
- reducing labor costs;
- accurate measurement and control of variables such as light, humidity, and water throughout the control year;
- increasing the reliable harvest of food;
- reducing water and energy consumption;
- reducing water consumption by up to 70% compared to traditional farms;
- use of robots for harvesting, planting, and logistics.

1.8.2 FARM AUTOMATION

Smart farm automation is a technology that makes farms more efficient and automates the crop or livestock production cycle. Every day, more and more companies are working on robotics technology to produce drones, autonomous tractors, robotic cars, and automated watering and seeding robots. Although these technologies are relatively new, the industry has seen some traditional companies enter farm automation. Technological advances, from robotics and drones to computer vision software, have revolutionized modern agriculture. The primary purpose of farm automation technology is to cover simple and mundane tasks. Farm automation technology addresses significant global population growth, farm labor shortages, and changing consumer preferences. Some of the leading technologies commonly used by farms are:

- harvest automation;
- autonomous tractors;
- seeding and removing weeds;
- drones.

1.8.3   LIVESTOCK BREEDING TECHNOLOGY

Another important industry that is usually ignored and has fewer services is animal husbandry. This essential industry provides renewable natural resources. Its management is traditionally known as poultry farms, dairy farms, cattle ranches, or other livestock-related agribusinesses. Managers must keep accurate financial records, supervise workers, and take proper animal care and nutrition seriously. Recent trends have proven many technologies have been developed in the world of livestock management and have led to significant advancements in the industry. This way, livestock tracking, and management are much more manageable and based on data. Livestock technology increases the productivity, welfare, or management capacity of animals and livestock. In the meantime, the concept of "connected cow," which means equipping herds with sensors to monitor health and increase productivity, is introduced. Placing these individual sensors on cows can track daily activity and health issues.

At the same time, data is also collected. All this generated data is transformed into meaningful and actionable insight, and producers can make quick management decisions with a quick and easy glance. Meanwhile, there is also a discussion of animal genes and their study. Animal genomics helps livestock producers understand the genetic risks of their herds and determine future profitability. Cattle genomics enables producers to optimize herd profitability and performance by strategizing livestock selection and reproduction decisions.

1.8.4   MODERN GREENHOUSES, AND MODERN AGRICULTURAL TOOLS

In recent decades, the greenhouse industry has been small scale and used mainly for research and aesthetic purposes (e.g., botanical gardens). On the other hand, producers are looking for much larger spaces to produce food, usually on traditional agricultural land. It is interesting to know that the greenhouse industry in the whole world is approximately 350 billion dollars annually, and the share of greenhouse and modern agriculture in America is less than 1%. The share of modern Dutch agriculture is probably much higher. Greenhouses have made significant progress in recent years, with the market growing significantly. Modern greenhouses have improved the growing environment by using LED lights and automatic control systems. Growing greenhouse companies usually locate their facilities near urban centers to take advantage of the location of people and their ever-increasing demand for more sales.

1.8.5 PRECISION FARMING

New agricultural companies are developing technologies that allow farmers to maximize crop yields by controlling each crop variable, such as moisture levels, pests, soil conditions, and microclimates. By developing more precise techniques for planting and growing crops, farmers can increase efficiency and manage costs better. Precision agriculture companies have found an enormous opportunity for growth. A recent report by Grand View Research predicts that the precision agriculture market will reach $43.4 billion by 2025. If this kind of technology can be implemented in Iran, the import of vegetable food will probably decrease significantly.

1.8.6 BLOCKCHAIN IN MODERN AGRICULTURE

Blockchain capabilities track ownership records, resist tampering, and solve urgent issues such as food fraud, safety recalls, supply chain inefficiencies, and food traceability in the current food system. Blockchain's unique decentralized structure ensures proven products and methods for creating superior products. Traceability is critical to the food supply chain. On the other hand, the existing communication framework in the food ecosystem has made traceability a time-consuming task because some of the involved parties still track information on paper like before. But the blockchain structure ensures that every link along the food value chain creates and securely shares data points to create an accountable and traceable system.

1.8.7 ARTIFICIAL INTELLIGENCE IN MODERN AGRICULTURE

The emergence of digital agriculture and related technologies has created new opportunities for data. Remote sensors, satellites, and drones can collect information 24 hours a day over the entire earth. These items control plant health, soil condition, temperature, humidity, etc. The amount of data these sensors generate is enormous, and the importance of numbers is hidden in the pile of data. This idea allows farmers to understand the land's condition better and obtain more information about the land and plants through advanced technology (e.g., remote sensing). Information that cannot be seen with the naked eye will be available quickly.

Remote sensors enable algorithms to interpret a field's environment as understandable and valuable statistical data for farmers to make better decisions. Algorithms process data, adapt it based on incoming data, and even learn. The more input and statistical information, the better the algorithm predicts a broader range of outcomes. The goal is that farmers can achieve their goal of better harvest by using artificial intelligence and through better decision making.

1.9 DISCUSSION AND CONCLUSION

In today's world, when diseases are controlled and the population increases more and more, the importance of using these technologies increases. In addition, the benefits of this type of agriculture encourage gardeners and farmers to use it. So, in the not-too-distant future, we will witness the growth of these methods in our world. With the advancement of technology in agriculture and the growth of modern tools, it can be said that this industry will experience tremendous growth over the next two decades. With the information that modern tools give you about soil, flowers and plants, weather conditions, etc., the error rate is significantly reduced, and all products are harvested entirely and ripe. Making the initial versions of these tools is not the end of the work, and it should be seen how their progress will be in the future and what other technological tools will be made for this industry.

2 Agricultural Autonomous Vehicles (AAV) and Components Description

2.1 INTRODUCTION

Autonomous agricultural vehicles (AAV), as intelligent mechanical systems for agricultural, horticultural, and many other applications, consist of many different components. The development of an AAV requires sufficient knowledge about each part, its characteristics, applications, datasheets, and methodologies. Also, intermediate knowledge is necessary to combine several elements and apply the desired application. In this chapter, we will briefly introduce the primary classification for the components of AAVs, and then we will discuss each section in detail in separate chapters. It is recommended to read this chapter carefully because it draws a comprehensive generalities image of the AAVs. Also, this chapter will be the basis for studying the details of the following chapters. Adequate understanding and analysis of this short chapter will be the prelude to understanding the whole book.

Before describing the central systems, let's define three key words used frequently in this book: automation, mechatronic, and robotic. We believe these words are defined in different references, but we will describe them differently. Undoubtedly, we use other systems to facilitate our work progress in our daily life, professions, hobby, and many aspects of our life. In most cases, we select the more practical, user-friendly, and cost-efficient system. We don't use a robotic arm or an electric saw for our nail cuts. We need to know which of those systems with its characteristic can help us do our required task. Each method has different structures, prices, complexity, and other technical and economic factors. If we do not achieve the target structure before implementing a project, we may waste our budget and time and pass the deadline. For example, we never use expensive robotic arms to close soda bottle caps. We never apply a complicated learnable controlling logic for an automatic garage door. From another perspective, we do not expect complex behavior from a simple system such as an automated toy box labeling system. Let's define the automatic, mechatronic, and robotic systems from the intelligence and learnability point of view.

Let's start with the "automatic" system, repetitive activity that a device can do many times. The word "automatic" presents a machine or apparatus that operates automatically. An "automatic system" is a mechanical system or mechanism

DOI: 10.1201/9781003296898-2

equipped with sensor(s), actuator(s), and controlling unit (mostly one as its ECU[1] or CU[2]), which is designed to do a specific or a limited number of tasks for infinite times. The automatic systems do the task(s) many times. The sensor detects a sense, sends the data to CU, and based on the program written in CU, the command is sent to an actuator, and this task repeats and repeats. Suppose something happens in the course of this mission. In that case, these systems will only be able to identify the situation and will not be able to make decisions according to the new conditions. For example, when the fibers are stuck in a part of a wool spinning line, the system gives an error, stops moving, and waits for the operator to fix the error. The same happens in commercial printers, regular automobile brake alarms, and many other devices we use daily. Most production lines for different industries include many automatic system units that work in parallel, series, parallel-series, or complex. Each unit is designed to do some specific task (mission) continuously. One of the main advantages of an automatic system is its fast operation. For example, in a mineral water production line, many automated systems for washing, drying, filling, labeling, and packing units can produce 15,000 bottles per hour. As another example, there are many automatic units for washing, slicing, drying, and packing in an automatic fruit and vegetable drying product line. Each unit mentioned in the examples is known as an automated system. The systems that not make decisions in unforeseen circumstances and only activate the operator in the defined state if the sensor information is received.

In the second stage, there are "mechatronic" systems. As a simple description, a mechatronic system is an automatic system designed for many predefined situations. As the number of work conditions and scenarios increases, the speed of operation decreases. Because the system consumes a longer time to define the case, decide the operation scenario, and then act. A mechatronic system combines mechanical, electronic, control, and computers in many references. This system has many applications in aerospace, medicine, material processing, manufacturing, defense system, and many others. Imagine you, as an R&D researcher/developer, want to innovate a system or develop a system for some task. Your system needs to receive information as input, process it, and send commands to outputs. If the intake of this system is predefined for various conditions, the system is a mechatronic system. For example, as a simplified tomato sorting machine, the system not only can classify ripe tomatoes but also should sort unripe, damaged, spoiled, and small fruits. In this and similar systems, the inputs of the system and related acts are predefined using equations, graphs, or algorithms. The system compiles commands based on predefined inputs.

Finally, we reach the robotic systems. In simple words, a robotic system is a trainable mechatronic. In the beginning steps of developing a robotic system, the system is more similar to a mechatronic system with an extensive and critical difference: its trainable controlling logic. During operation, if a robotic system meets a new unpredicted condition, the logic helps to decide a better decision to act. In this system, the input is mostly not predetermined. The system must use different sensors to sense the environment and conditions and export the

TABLE 2.1

Characteristics Comparison of Different Smart Systems

Characteristic	Type of System		
	Automatic	Mechatronic	Robotic
Complexity	Medium or low	Medium or low	High
Application	Medium or low	Medium or low	High
Controlling system	Medium or low	Medium or low	High
Cost	Low	Medium	High
Number of sensors	Low	Medium	High
Controlling algorithm	Medium or low	Medium	High
Learnability	Medium or low	Medium or low	High
User friendly	High	Medium or low	Medium or low
Environmentally friendly	Low	Medium	Adoptable
Artificial intelligence application	Low	Medium	High
Energy using	High	Medium	Adoptable
Work speed	High	Medium	Medium or low

appropriate command using its control algorithm. For example, a self-driving unmanned vehicle uses various sensors (leader, laser scanner, camera, positing sensor, orientations sensors, etc.) to examine the environment and related simulation and analysis, determining which direction and speed must move. The robotic systems are designed for a (mostly) unpredictable environment that has many variables, and the robotic systems must provide the best performance. Some examples are the auto-driven Tesla cars, Boston dynamic robots, and Yanmar autonomous tractors. Many different logics, such as PID[3], MPC[4], Fuzzy logic, or ANN,[5] can be used in robotic systems. Table 2.1 compares the characteristic of mentioned systems.

2.2 AGRICULTURAL ROBOTS

Agricultural robotics are being developed to revolutionize the agricultural industry. In addition to transforming, these technologies must reduce the disadvantages of existing technologies in agriculture. These technologies can be most effective in areas with relatively low productivity [72]. Agricultural robotics is one of the most important subdivisions of digital agriculture, which has been able to integrate a variety of modern technologies with a variety of equipment and algorithms practically and scientifically so that it can turn tedious operations into continuous automated processes [73]. Applications of agricultural robots include growing plants, fruits, and vegetables in farms, gardens, and greenhouses. According to the rules based on precision agriculture and data science, this tool accelerates the growth process, reduces energy costs and consumption, and even reduces production inputs such as labor and equipment [74]. This emerging technology

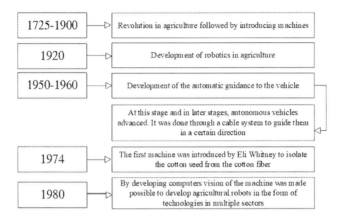

FIGURE 2.1   The timeline of agricultural robots and their progress.

reduces duplication of work in agricultural operations and, on the other hand, increases crop productivity by monitoring each crop separately, which is also a subset of precision agriculture [75]. In the early 1920s, the first step was to develop an automated vehicle navigation system. In agricultural applications, the first driverless tractors were developed in the 1950s and 1960s, with cable-guided navigation systems [75]. Figure 2.1 presents the progress of agricultural robotics.

Then there were fundamental changes in agricultural robots to harvest oranges. Due to their high complexity, agricultural robots were primarily used indoors and outdoors. The use of these robots has uncertainty from a security point of view, picking agricultural products and presence in various environmental conditions that entirely affect the system's performance. Numerous studies have been conducted to investigate robotic technologies' state-of-the-art in agriculture in various applications. This section briefly mentions some of the state-of-the-art studies in agriculture robotics.

Saleem et al. (2021) prepared a survey study to evaluate the performance of agricultural robots developed by ML and DL techniques. According to the findings, a region-based convolutional neural network has been proposed for plant disease/pest detection rate [76]. Zahid et al. (2021) investigated a robotic pruner's commercial and scientific progress. The applications include machine vision, manipulator, end-effector, path planning, and obstacle avoidance. In the following, the challenges and potential solutions associated with these technologies have been discussed [77]. Lytridis et al. (2021) presented comprehensive state-of-the-art technologies for cooperative agricultural robotics. This paper also extracts the challenges identified in fully automating agricultural production [78]. Atefi et al. (2021) discussed the developments in phenotyping robots. In addition, the appearing challenges have been considered. In the following, the future perspectives and applications of phenotyping robots have been accessed [79]. Martin (2021) presented future rural regulations for accessing future agricultural innovation. The study presented a careful consideration to examine opportunities to cope with agricultural robotics. Policy-makers can successfully employ the main findings of the study for agricultural sectors and innovation industries [80]. Oliveira et al. (2021) presented a comprehensive study discussing recent advances in agricultural robotics for land preparation before planting, sowing, planting, plant treatment, harvesting, yield estimation, and phenotyping [81]. Fountas et al. (2020) developed a systematic survey to describe research and commercial agricultural robotics employed for crop field operations. According to the findings, harvesting and weeding have the most employed robotic systems, and the lowest share is related to disease detection and seeding robots [82]. Ren et al. (2020) developed a review work to emphasize robotics enabling machine capabilities and their technical challenges [83]. Vougioukas (2019) developed a study to reflect the highlighted points on ground robots in agricultural environments. The findings addressed challenges and future perspectives along with their limitations [74]. Zhao et al. (2016) presented a comprehensive review to discuss recent advances in vision-based control systems handling fruit or vegetable harvesting robots. In the following, the study discussed vision control techniques and their potential applications for the future challenges [84].

2.3 THE OVERALL CONFIGURATION OF AUTONOMOUS AGRICULTURAL VEHICLES

An autonomous agricultural vehicle is a mechanical platform with actuators, control units, sensors, and related algorithms. It maneuvers in a target land (farm, orchard, rice paddy, and road) to do one/various predetermined tasks (such as plowing, carrying objects and equipment, harvesting, watering, weeding, seeding, seedling, etc.). Each part is shown in a separate segment in Figure 2.2. The components of an AAV work in interaction with each other. In the following sections and chapters, each segment will be explained comprehensively.

2.3.1 PLATFORMS AND MECHANICAL COMPONENTS

In this book, the platform and mechanical components of an AAV are interpreted in Chapter 3. The platform is a unified mechanical structure; all components are installed. The outline of Chapter 3 is as follows:

- Different development procedures (prototype, semi-commercialized, and commercialized)

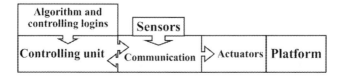

FIGURE 2.2 Configuration of autonomous agricultural vehicles and their components.

- Different platforms of an AAV (tractors, combine harvesters, transplanters, utility vehicles (UTVs) and all-terrain vehicles (ATVs), drones, and prototype platforms)
- Different transportation systems (tires, crawler and half-crawlers, robotic foot, and propellers)
- Different actuators and motion functions (steering, brake, shift, hitches, and PTO)
- Different power sources (vehicles with an internal combustion engine (ICE), vehicles with electric motors, battery electric vehicle (BEV), hybrid electric vehicle (HEV), plug-in hybrid electric vehicle (PHEV), and fuel cell electric vehicle (FCEV))
- Different power transmission

2.3.2 SENSORS

The different required sensors to maneuver an AAV are presented in Chapter 4. In an AAV, sensors are electrical/semi-mechanical/mechanical devices that can detect various information as digital/analog inputs from the operating environment. In agricultural environments, the inputs can be obstacles, positions, attitudes, and many operation functions. The sensors send a signal which is readable to the controlling unit. The outline of Chapter 4 is as follows:

- Different positioning sensors (navigation using a video sensor, lateral navigation (LNAV), X-ray pulsar-based navigation (XNAV), global positioning system (GPS), smooth navigation (SNAV), differential GPS (DGPS), and real-time kinematic-GPS (RTK-GPS))
- Different orientation sensors (geomagnetic direction sensors (GDS), fiber optic gyroscopes (FOGs), inertial measurement units (IMUs))
- Different safety sensors (wave-based safety sensor, vision-based safety sensor, sound-based safety sensor, and contact-based safety sensor)

2.3.3 COMMUNICATION

After that, the different communication methods between components of an AAV are presented in Chapter 5. In AAVs, the communication methods are different ways different electronic components communicate. Different communication methods differ in speed and quality of data transmission, wired or wireless, and various parameters. The outline of Chapter 5 is as follows:

- Internal communication methods (direct control systems, serial communication, and CAN-bus)
- External communication methods (Wi-Fi, Bluetooth, infrared, and LTE)
- Different types of ports and their characteristics (parallel port, RS port, PS/2 port, audio port, DVI port, HDMI port, VGA port, USB port, Ethernet port, S/PDIF/TOSLINK port, SATA port, and Apple ports)

2.3.4　Controlling Unit

Different controlling units of AAVs are presented in Chapter 6. As an AAV's brain, the controlling units play a crucial role in receiving data from sensors, making decisions based on the algorithms, and implementing commands by actuators. The outline of this chapter is as follows:

- Different controlling units (controlling units (controlling boards (Arduino, Raspberry Pi, BeagleBone Black, and AdaFruit)) and computers (microcomputers (personal computers), mini computer, mainframe computer, and supercomputer)
- Performance comparison of the controlling units

2.3.5　Algorithm and Controlling Logins

The algorithms and controlling logic of AAVs are presented in Chapter 7. This section is one of the essential parts of this book because, in different environments, transportation, and different segments of operations, different parameters affect the control of AAVs. Also, controlling logic has been used to control a vehicle over time, and based on the academic reports, they have shown different performances in different situations. The outline of this chapter is as follows:

- Main performance indicators: heading and lateral error
- Evaluation indicators (accuracy recall precision, RMSE, determination coefficient)
- Different controlling logics (artificial neural networks (ANNs), fuzzy logic, proportional, integral, and derivative (PID), model predictive control (MPC))

2.3.6　Software and Apps for Design

All developments required various software and apps in different aspects. Chapter 9 outlined the most essential and applicable software and apps as follows:

- Software and apps for controlling (Myrobotlab, Microsoft visual studio, Processing, Notepad++, and ArduinoDroid)
- Software and apps for performance simulation (simulators) (robot operating system (ROS), and Gazebo)
- Software and apps for design (design and simulation) (Autodesk Inventor, DraftSight, SolidWorks, AutoCAD, OpenSCAD, CATIA, and SketchUp)
- Software and apps for circuit development (Fritzing, CadSoft Software Eagle PCB Design 95, Falstad Circuit Simulator, DipTrace, KiCad EDA Software Suite, gEDA, OpenCV, TINA, and Eagle for board layouts)

NOTES

1 Electronic control unit.
2 Controlling unit.
3 Proportional–Integral–Derivative.
4 Model Predictive Control.
5 Artificial Neural Network.

3 Platforms and Mechanical Components

3.1 INTRODUCTION

One of the main parts of any AAV is its platform and mechanical components. These parts of a robot contain all the visible and physical parts except electrical components and control units. In other words, all electronic components, control units, communication systems, and other parts attach to mechanical structures, which we named "platforms." As shown in Figure 3.1, this is one of the earliest parts of AAV development that researchers should consider. As mentioned, the platform is a mechanical structure with all components installed. For example, an autonomous car is a set of intelligent systems attached to a car. In this example, the car is the platform of the robot. An unmanned aerial vehicle is a drone with an auto-navigation system. Literature, the platform, is the backbone of a robot. This part is essential because a perfect selection of platform and mechanical components directly impacts your robot's performance. Imagine you select a wheel-type UTV to transport firewood in the depth of a forest.

The transportation system (wheel-type in this example) will directly decrease the performance of your system while overcoming obstacles. Or if you want to apply autonomous navigation on an old-type tractor that has a manual gearbox and mechanical steering system. In this case, you will struggle with numerous challenges in controlling this system using your controlling unit. Some of the researchers in this field pay less attention to this aspect. After completing most of the project, they look for solutions to related problems. To avoid dealing with similar problems in your development project, we will first discuss platforms and mechanical components, their types, characteristics, principle, and all required information. In this chapter, we will first discuss AAV's different development procedures. An optimized development procedure can help you to speed up your project execution. Then, different platforms of an AAV, including tractors, combine harvesters, transplanters, utility vehicles, all-terrain vehicles, drones, and prototype platforms, plus their types and examples. More on that, we will focus on more details such as transportation systems (wheel-type, crawlers, half-crawlers, feet, and propellers), actuators, and operation functions (steering, brake, forward and backward, shift, rotary speed, hitch, and PTO), power sources (internal combustion engine (ICE), electrical motor, hybrid source, and fuel cell power sources), and power transmission. Each element will be discussed and detailed.

DOI: 10.1201/9781003296898-3

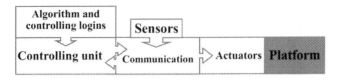

FIGURE 3.1   Platform in the autonomous agricultural vehicles.

3.2   DEVELOPMENT PROCEDURE

Firstly, let us discuss the development process. All successful systems, including agricultural robots, have passed through a development process. When you see a farming robot, be sure that the robot had a proud birth in a laboratory. From brainstorming, idea generation, synthesis, sorting, labeling, best selection, and prototype development. However, the idea passed through an ideation and qualification process; the experimentation results can push you to repeat the ideation process again and again. In this step, your accurate and wise friend and colleague can help you by objecting to your proposed idea. One of the cost-effective ways to deal with your future financial/technical failures is to identify problems at this point. If your idea is to fail, it's best to fail as soon as possible. Indeed, the idea that came from the heart of several failures and corrections is more effective. After passing the idea generation stage, the system will develop in a laboratory or R&D[1] section of a company, which we name a "prototype" system. At this stage, the structure is made regardless of its beauty. The prototype design and even the system made from it are not the best to offer to the market. So before entering the market, it must undergo the development process.

As shown in Figure 3.1, the beginning of the development process is designing and evaluating the robot in a laboratory. The prototype system is subjected to various tests several times to make sure that the system can enter the market as a semi-commercialized system. After the final evaluation success, the semi-commercialized system is almost the same as a prototype system with more consideration for its beauty and visual effects. Before the semi-commercialized stage, researchers have the opportunity to address all the problems that can happen in the system. If you are a beginning designer of an AAV system, relax! You can often make mistakes in selecting parts, designing, and even programming. You can do many modifications and finally provide a presentable system to the company or even the market. This stage's outcome must be tested in the real world and exceed the required standards.

In the semi-commercialized stage, before presenting the system on the market, the selective customers or examiners find problems with its use or components. In this case, the manufacturer and the designer researcher must re-test the system and fix it. Sometimes solving the issues is not easy, and significant changes are required. The AAV should be transferred to the laboratory to redesign the system with newly defined parameters. As you can see in the Figure 3.2, the designed robot is often exchanged several times between the prototype stage and the semi-commercialized stage to achieve an acceptable performance. Finally, the

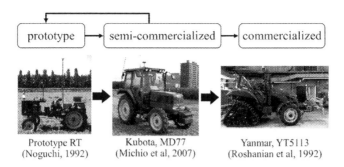

FIGURE 3.2   Development process of an AAV.

development process enters its final step: a "commercialized" product. At this stage, the commercialized system could pass all qualifications and standardizations.

A tractor and its controllable function can be a practical example of this section. At first, the tractor may not have all the components required as an AAV (Figure 3.2). In this case, researchers in the laboratory and factory try to modify various parts such as mechanical components, actuators, chassis, and wheels. The modification is developing a platform that can be used for robotic applications. The previous tractors had no central control or an automatic gearbox. In this case, a laboratory may request developing and installing an ECU and automatic gearbox suitable for farm use. At this point, the output of the modification can be converted into a tractor with extra wheels, a central control system, a controllable PTO, a controllable steering wheel, or even different types of wheels. The Yanmar tractor (model: YT5113, Japan) is an example of a commercialized platform suitable for autonomous tractors. This platform does not require mechanical modifications to apply controlling systems, algorithms, etc. This platform has a connectable and programmable ECU through various communication methods.

In conclusion, the prototype system is developed in the development process of a platform. It reaches the semi-commercialized stage using the necessary modifications, and finally, it is considered a commercialized platform. Researchers in developed countries such as Japan have the task of modifying to develop a robotic system, and the manufacturer does the modification. It is good to mention that the ideal commercialized platform for robotic systems should have a controllable steering wheel, brake, clutch, ECU, automatic gearbox, and other actuators. The ECU interacts with the central computer of the system.

3.3   INSPIRATION FROM NATURE FOR DESIGN

One of the surest ways to design a new system for a specific application is to take inspiration from similar structures. Nature has extensive structures that have been developed over the years. The development of natural structures is based on mutual needs, creating confidence that modeling these structures can be trusted. One of the processes of adapting from nature is to extract the required activity before taking each step. A goal can help find the best biological structure

TABLE 3.1

Examples of Inspiration of Nature to Design AAVs

Required Action	Aim	Example of Application	Inspiration Source	Ref.
Complicated motion	Moving in rocky and rugged terrain	Legged robots in fields	The human leg, animal leg	[85,86]
Limbless locomotion	Access to complex location	Flexible robotic arm for apple harvesting	Worm, snail, caterpillar, snake	[87]
Climbing	Climbing at a sharp angle	Climbing a tree to cut branches	Gecko, ant, goat	[88]
Stickiness	Adhesion to reduce damage to products during harvest and transportation	Harvesting fragile fruits	Gecko	[88]
Jumping	Product transfer	Jumper robots	Bharal, hare, kangaroo, grasshopper, flea, and locust	[89,90]
Floating	Swimming in liquid environments	Airboat in paddy field	Fish	[91]
Moving in a dense group	Using collective intelligence and dividing tasks	Multi-robot tractor	Ant	[92]
Grasping and moving	Selective harvesting	Pumpkin harvesting	Elephant	[93]

according to the need. After identifying the biological structure, the next step is redesigning and analyzing the relevant structure. After this stage, the design is evaluated and optimized if necessary until the required application is obtained. Table 3.1 presents a set of examples related to this topic. You can inspire ant group work to design a multi-robot tractor, an elephant trunk to design a manipulation system for heavy-duty, caterpillars to climb on trees, or kangaroos to jump through crops during loading. Believe that nature is a rich source of inspiration that gives you the idea to design innovative systems.

3.4   PLATFORM

The robot platform is an integrated mechanical structure, including the body, chassis, cabin, etc., on which the rest of the farmer's robot parts are installed. We will study the device on which all the robotic equipment is mounted in the platform of this book. The tractor and the combine are the primary tools for working on the farm and gardens. Other tools used in agriculture include transplanters, UTVs, ATVs, boats, and aerial vehicles. In the coming sections, we will discuss each of them in detail. Also, these sections will explain their advantages, disadvantages, and capability to be robotized.

3.4.1 TRACTOR

Since the tractor is the oldest tool for agriculture, it should have a larger share of future robots. This platform (more modern versions) has the highest capability and chance for smartening. This device is used for almost all stages of agriculture, including planting, transportation, harvesting, landscaping, seeding, fertilizing, irrigation, spraying, and many more. The advantage of tractors in their present form began about 150 years ago on English farms, which were used instead of horse traction. Prototypes of tractors were steam-powered plows. This steam engine needed large amounts of water or coal to work; therefore, these early tractors were tremendous and required a lot of care and maintenance, so sometimes 15 workers were needed to move and use a steam engine tractor. These early tractor models had huge metal wheels that could support the hefty weight of the tractor. Of course, the metal of the tractor wheels reduced the car's speed. However, these tractors were standard until the early 20th century. Later in 1940, hydraulic mechanisms were added to load and increase power and controllability after the invention of the hydraulic system. In a general classification, tractors are divided based on their application. Different types of tractors (row crop, general purpose, tracklayers, half crawlers, and two-wheeled) are presented here.

3.4.1.1 Row Crop Tractor

These tractors are small tractors designed to move in the cultivation rows for horticultural and agricultural lands. Due to their small size and agility are mainly used for growing vegetables and root crops (hooch, beet, etc.) (Figure 3.3). Depending on the application and size, they are agile and have a small rotational radius. A smaller rotational radius reduces the length of the steering wheel area at the ground end, increasing work efficiency. Due to their low cost, they are one of the recommended platforms to start the agricultural automation and intelligence project. As their power range is low, they need extra batteries for heavy-duty and external equipment installation, such as computers, sensors, and electrical devices. This platform is suitable for testing and experimenting with a new robotic project. It is because they are cheap (low cost in case of breakdowns and necessary repairs), small (to check the performance of the designed system and evaluate it in smaller spaces), and easy to control (during evaluation and due to the simplicity of different sections). These tractors have more accessible settings and easier control, which makes them easier to program. As the length and complexity of the written program decrease, it requires a lower-power computer system. As a result, the weight of the extra batteries to power the systems decreases.

On the other hand, the simplicity of the system and more straightforward programming will increase the speed of program execution and increase the accuracy of robot performance. These tractors provide a complete enclosure for the driver to monitor the cultivation rows. They have a small size with an

FIGURE 3.3   An example of a row crop tractor (Yanmar, Model: EG453 [Yanmar Co., Japan]).

approximate power of 15–60 kW, which is suitable for light applications. In the old systems, these tractors had diesel engines with three cylinders; today, they mostly have four cylinders. In recent years, due to the lower weight of these tractors, their less negative impact on soil compaction, and their high flexibility to define team-robotic activities, such as multi-robot tractor systems, have been accepted by researchers in this field. If you have a futuristic design, this platform is the best choice for developing a multi-robotic system based on dense group work.

3.4.1.2   General Purpose Tractor

These tractors are the large tractors we see most of the time on big farms. This tractor has an engine power of 60–150 kW, a driver's cabin, and is extensive (Figure 3.4). Due to the high power and high torque they produce, farmers prefer to use these tractors for various tasks. These tractors are more complex than the previous group because more power is required for more complete parts. For example, the hydraulic system of a tractor may be much simpler and more repairable for cultivation rows, but tractors in this group have more complicated settings. The performance of these tractors has been controlled using ECU. Depending on the size and type, these tractors can be used for various robotic applications. But it should not be forgotten that as the size of these tractors increases, the ground pressure increases too. This action will cause soil compaction. Although traffic can be reduced in hybrid systems, the high weight of hybrid devices can cause condensation.

FIGURE 3.4   An example of a general-purpose tractor (Yanmar, Model: EG106 [Yanmar Co., Japan]).

This type of tractor platform is very effective for large farms. The ability to pull and lift heavy objects is one of the main characteristics of these tractors. They have a powerful hydraulic system and a more powerful PTO. It should be noted that to choose this platform as a reliable platform for robotic applications, it must have a standard steering system, gear shift, brakes, and throttle. Most modern tractors in this group have four active wheels. In appearance, almost all tractors in this group have toothed wheels. In rare cases and for specific applications, this group has powerful tractors with 250–390 kW (340–530 hp) and a maximum speed of 30 mph. Although it can be said that if you use these platforms for your designed robotic systems, you must have enough reasons to use such a heavy platform for your robot. It is because the future farmer robots are moving towards more miniature robots.

3.4.1.3   Tracklayer

The difference between these tractors and the last two groups is their wheel type, not the engine power. These tractors are designed to reduce pressure on the soil. Since the contact surface of the crawler is much bigger than the wheel type, assuming the weight of the tractor is constant, wheels apply more pressure to the ground than the crawler (Figure 3.5). Today, most farmers use active four-wheel tractors to reduce the risk of erosion, floods, and soil compaction. These types of wheels are either metal or rubber. Metal crawlers are more durable, but high depreciation and not being allowed on the road are the disadvantages of this type of crawler. The rubber crawler, one of the newest types, can be used on agricultural

FIGURE 3.5   An example of a tracklayer (Yanmar, Model: CT801 [Yanmar Co., Japan]).

land and roads. They produce less noise and have less wear and tear. Since crawlers require a more complex design, they have more parts and higher maintenance costs. If you want to use these tractors as your robot platform, you should note that these tractors are slower than general-purpose tractors. Due to more dentin and the higher contact surface, the structure meets significant vibration on the road surface. If you do not use an SSD hard disk for your robot's computer, it will likely be damaged after a few hours of operation. This vibration can also damage other parts of the tractor. But it is interesting to know that the vibration of these tractors in agricultural lands is less than the previous two groups.

3.4.1.4   Two-Wheel Tractor

Finally, the last group of tractors is called two-wheeled tractors. These tractors often have one or two cylinders and use two-stroke engines in most cases. They have two wheels mounted on one axle (Figure 3.6). The direct force of the hand does the rotation of the steering wheel. This device has a steering wheel similar to a motorcycle steering wheel, it is steered manually, and the driver or operator walks after it or sits on the vehicle that is closed behind it. The tillers have a PTO output that supplies the necessary power to the connected devices. The devices that attach to the motor rotator include wheels to move the machine, rotary blades, earthing wheels, plows, and harnesses, which are usually traction. Although this device does not have many talents for becoming a robot, no researcher has tried to make this device smart.

FIGURE 3.6   Examples of a two-wheeled tractor.

3.4.2 COMBINE HARVESTER

Combines are the second most widely used platform for robotic systems in agriculture (Figure 3.7). These tools are used for harvesting agricultural products, mainly grain crops such as wheat, rice, barley, corn, and soybeans. In brief, this platform performs three primary operations: harvesting, threshing, and separating. After the invention of this device, the dependence of consultants on the labor force was significantly reduced, so with the help of this device; it is possible to have at least one worker cut the product. Although the internal system of combine harvesters is more complex than tractors, and many adjustments must be made to perform the harvesting operation, driving and managing newly produced combine harvesters is very convenient. It would help if you had a joystick (direction and speed control) and a brake pedal to control them. This feature makes this device highly capable of becoming a robot. It should be noted that due to the structure of this device, many physical parameters should be included in the control algorithm. Still, we will not have many challenges with its robotic control. Examples of combine harvesters are John Deere, Model: CH570 for cotton; John Deere, Model: CP690; John Deere, Model: T670I (John Deer Co, USA); and Yanmar, Model: YH590 (Yanmar Co., Japan) for sugarcane, cotton, and cereals harvesting.

FIGURE 3.7   Examples of combine harvesters.

FIGURE 3.8   Examples of rice and tomato seedlings transplanter.

3.4.3 TRANSPLANTER

Transplantation is one of the methods to accelerate cultivation, control weeds, and reduce pests' risks. Transplanters input seedlings planted directly in the ground (Figure 3.8). Various seedlings for planting crops are rice, tomatoes, and vegetables. In most countries, such as Iran, some crops, such as rice, are planted using seedlings. Transplantation usually consists of two main parts: the stimulus or stimulants and the transplanting system. The seedlings behind the tractor use a tractor as the propulsion system. If you want to design a robotized transplanter, it is necessary to pay attention to the purpose of the study. All the points mentioned in the tractor must be followed. But to make the transplant sector smarter, it is necessary to pay attention to the purpose of changing the mechanism or making this section smart. Suppose the efficiency of the designed system is not sufficient. In that case, we can try to improve this part by performing mechanical analysis, but if it is necessary to study the performance and create a transplanting algorithm based on different sensors, the operators of this part should receive commands from the central control system. It is clear that to send the necessary commands the required data must be received from the relevant sensors connected to the components of this player.

On the other hand, some seedlings are semi-automatic and require the presence of one person to place the seedlings in the path of the planter mechanism. One of the capabilities and opportunities for smartening this platform is to create a fully automatic system and replace the essential systems with the labor force. In most of the articles presented on this platform, research has focused on the device control section. The main goal of this platform was to make the transplant smarter. Since transplanting on some products can increase production repercussions, it is proposed to design and implement more efficient and intelligent mechanisms for this activity.

3.4.4 UTILITY VEHICLES (UTVS) AND ALL-TERRAIN VEHICLES (ATVS)

More people believe tractors are the only vehicle to transport in the field. Tractors are significant; they use more fuel, and their maneuver is complicated.

Except for the main farming operation, tractors are not the optimized vehicle in the field. Two types of vehicles, ATVs and UTVs, were invented for such purposes. These multi-user tools are smaller motor tools than tractors that are usually designed to do one or more specific tasks with higher efficiency. These platforms are recommended for robotic projects because of their modification capabilities, smaller size, and agility.

ATV means a "vehicle for all applications" and is a full-fledged and all-encompassing vehicle, also informally known as a four-wheeler or four-wheeled motorcycle (Figure 3.9). According to the American National Institute of Standards, it is a vehicle that drives on low-pressure wheels, making it suitable for agricultural ancillary work and platform robotics. This platform has a seat for the driver, and its steering wheel is of the motorcycle steering type, which can be controlled by installing a rear servo motor. As its name suggests, this vehicle can move in various areas and surfaces. It is also used in some countries as an off-road vehicle. It works like a motorcycle, and the difference is that it has less speed and more stability due to more wheels. ATVs can carry 56–180 kg and have a higher center of mass, which increases the probability of overturning. This device is very useful in off-road activities, especially on farms and gardens, because it has better maneuverability and a high ability to move through obstacles.

UTVs are passenger car–like vehicles with two-way seats, two-way inputs, steering wheel, wheels and pedals, seat belts, occupant protection, and trunk (Figure 3.10). But compared to passenger cars, they have a higher load capacity, are more prominent, and are more comprehensive than most. The carrying capacity of this device is 360–620 kg, and they have a lower mass center and higher endurance than ATVs. These devices are designed to carry heavy equipment that requires high tensile strength for loading. They create high friction and high ability when moving on soft surfaces. All the capabilities of this platform are for robotic systems, such as tractors, with the difference that it has a wide range of operations in the side volume.

These devices are used in agriculture, sports, and military applications. Applications of these devices include transporting heavy equipment, plowing fields, collecting grass and additional materials with a rake, leveling, breaking and hoisting the ground, snow removal, lawn mowing, fencing, seeding, assisting

FIGURE 3.9 Examples of ATVs.

FIGURE 3.10   Example of UTV.

in controlling animals during grazing, transporting firewood, and transporting small volumes of soil.

3.4.5 DRONE

It may seem strange that we have placed drones in the segmentation of agricultural platforms, but I must say that this device will be an integral part of modern agriculture shortly (Figure 3.11). It has a body, power supply system, central control, sensors, actuators, control algorithm, flight control, and communication

FIGURE 3.11   Examples of drones.

system. In modern agriculture, this device is used for accurate aerial photography, spraying, irrigation, fertilizing, and correcting position sensor data. Apart from agriculture, drones are now widely used in telecommunications, global navigation, and meteorological and geographical research. UAVs will play an essential and influential role in identifying and tracking farming robots in the future. Drones will be one of the cheapest and most feasible solutions for receiving high-resolution images, extracting information about soil and plants, and feeding information to farming robots that will be accurately engaged in farming. It is noteworthy that sometimes boats, such as spraying on rice paddies, are used. So, who can say that different boats (motor, electric) can also be platforms for agricultural robots?

3.4.6 PROTOTYPE PLATFORM

Most of the time, the design of an AAV starts with a controlling system applied on a prototype platform. The prototype platform is an uncommercialized mechanical base to develop a new robotic system because (1) the desired platform is not available in the market, (2) the commercialized platform is expensive or costly to use in the laboratory step of development, or (3) the designer wants to simplify the experimentation condition and, test/improve the system step by step. As shown in Figure 3.12, Noguchi and Ishii [94] developed an autonomous robot tractor on a prototype tractor. There are also many other prototype platforms, such as Acorn (an open-sourced platform for various applications [95]), cotton harvesting robots [96], or robots for mallard navigation in paddy fields [97].

FIGURE 3.12 Example of prototype platform (a prototype robot tractor [94]).

3.5 TRANSPORTATION SYSTEMS

The transportation system is another component of a farming robot that receives less attention in other disciplines. The transportation system is one of the robots' main components. In agricultural activities, the type of transportation system can significantly impact the performance of the designed robot due to the different movement levels of vehicles. Here, the transportation system is the part of the tractor robot that is in contact with the ground or causes the platform of a robot to move. More simply, the wheel in a robotic system is a transportation system.

As you can see in Figure 3.13, the moving units of the farmer robots are divided into five general types, (1) wheel-type, (2) half-crawler, (3) crawler, (4) robotic leg, (5) and propeller. Each type of transportation system is designed for different applications and functions, each with advantages and disadvantages. A designer should compare his goals with the advantages and disadvantages and choose the best type of transportation system for his desired application. For example, a platform with a wheel-type transportation system to work in paddy fields is not a good choice. Because of their physical behavior versus muddy soil, wheels are unsuitable for wet and swampy terrain. So if a designer is not familiar with the characteristics of each of these transportation systems, they may make a mistake in choosing it. As a result, the performance of the system will be questioned. Table 3.2 lists the features of each transportation system.

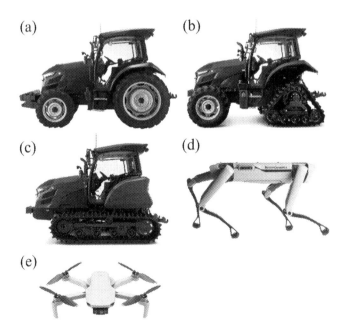

FIGURE 3.13 Types of transporter systems, (a) full-wheel, (b) half-crawler, (c) crawler [98], (d) robotic leg [99], and (e) propeller [100].

TABLE 3.2

Technical Comparison of Different Transportation Systems

Parameter	Wheel-Type	Half-Crawler	Crawler	Foot	Propeller
The complexity of designing and development	Low	Average	High	Complex	High
Reparation	Easy	Complex	Complex (easy to break)	Complex	Easy
Durability/Lifetime	High	Average	Low	Average	High
Cost/price	Low	High	High	High	Low
Material type	Various	Metal/rubber (sprocket) Metal (structure)	Metal/rubber (sprocket) Metal (structure)	Various	Various
Weight	Light	Average	Heavy	Average	Light
Ground pressure	High	Average	Low	High	N.A.
Soil compaction	High	Average	Low	High	N.A.
Power efficiency	Low	Good	High	Low	Good
Friction and adhesion	Low	High	High	Low	N.A.
Maneuverability	Good	bad	limited	High	High
Movable environments	Dry, solid, semi-wet	All surfaces	All surfaces	Dry and solid	On air
Move on sloping, slippery, and wet surfaces	Not good	Good	Good	Not good (depends on end-point)	N.A.
Drive over obstacles	Hard	Easy	Easy	Hard (depends on end-point)	N.A.
Traction system	Not-optimized	Optimized	Optimized	N.A.	N.A.
Commandability/ Steering	Good	Suitable (dry surfaces) Medium (wet surfaces)	Poor	Poor	Good
Rotate in place/Spinning	Good	Better than Wheel	Better than wheel	Good	Good
Turning space	Big	Average	Small	Average	Small
Speed	High	Average	Low	Average	High
Precision in movement	Good	Average	Bad	Average	Bad

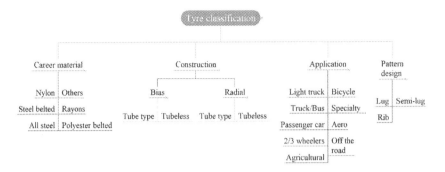

FIGURE 3.14   Tire classification.

3.5.1 Tire

The popular, repairable, and light-weight transportation system is all-wheel drives, wheels, or tires (see Figure 3.14 for tire classification). These transportation systems are economically viable for various applications, but they apply high pressure on the surfaces and cause soil compaction. As the contact surface of the tires decreases, the pressure on the soil and compaction increases. The quality of tires has improved in recent years, thanks to numerous advances in tire technology. Farmers travel longer distances with their equipment to access all their farms. Even modern hybrid corn with stiffer stems can damage expensive agricultural machinery tires. This problem in heavyweight tractors (above 100 hp) dramatically increases, which forces farmers to prevent repetitive movement on the farm for the long term. As a result, today's tires need to be improved to prevent soil damage. Table 3.3 presents the types of tires and their specifications, and Table 3.4 presents the types of agricultural tires.

3.5.2 Crawler and Half-Crawler

Another type of transportation system is a continuous chain or crawler. Crawlers have several wheels joined together by hard steel or rubber sheets. These transportation systems put less pressure on the surface and reduce soil compaction due to larger contact areas than wheels. This kind of transportation system makes moving on sloping and wet surfaces easy. Wheels on these surfaces get boxed because of low adhesion and often get stuck in the soil. Still, crawlers have reasons for their higher contact surface and different mechanical properties (which exert force on the soil surface in different directions). They can move on surfaces with different slopes and in swampy and waterlogged environments. However, it causes a lot of vibration in the tractor cab. Vibration is a significant parameter in the design of tractor robots. Vibrations and shocks are among the undeniable parameters important during agricultural operations. This parameter can negatively affect the performance of a farmer robot. Imagine designing a tractor; you need to install a computer in the cabin with vibration-sensitive components such as a hard drive. Note that you must use SSDs type

TABLE 3.3
Different Types of Tires and Technical Comparison

Type	Specification	Components	Function
Pneumatic: Tube type or tubeless design	Flexible and maintains its shape by air pressure	Hollow rubber	Radial-ply and standard on most cars today
Retreads tires	A method of saving money	Adhesive, rubber/chemical layer	Often used on trucks and buses
Semi-pneumatic Tires	Lightweight, puncture-proof, and provide cushioning	Not pressurized hollow tires	Wheels for lawnmowers, shopping carts, and wheelbarrows
Solid tires	Rigid, stiff, and durable for long-term work but with minimal impact absorption	Solid rubber, in combination with plastic compounds	Lawnmowers, scooters, and many types of light industrial vehicles, carts, trailers, material handling equipment (forklifts)
Cushion tires	Similar to a solid tire, a sealed internal air space	Layers of rubber or dense foam	For smooth surface applications, primarily indoors or on loading docks

TABLE 3.4
Different Types of Tires Are Used in Agriculture

Type	Application	Specification
Agricultural Tractor Tires	Fieldwork	The highest amount of traction
Turf Tractor Tires	Golf carts and lawnmowers	The lowest traction
Industrial Tractor Tires	Construction sites, especially on forklifts	Excellent traction and load durability

hard disks that are less affected by vibration. Also, the installation site of electronic performance should be protected against vibration in various ways, as well as dust and dirt. Other disadvantages of this transportation system are the diversity of parts, higher prices, limited speed, and difficult repairability compared versus wheels. When speed increases, vibration in the cabin also increases. It needs a shorter service life, has less maneuverability, and limited controllability.

These are more suitable for heavy vehicles such as combine harvesters and large general-purpose tractors. It causes road destruction due to its physical structure. It can cross hillsides and mud roads. Crawler tractors act with the help of caterpillars. Units with a mixed type of mover are driven simultaneously by wheels and caterpillars. On snowy roads, dirt roads, agricultural lands, sludge fields, sand fields, and dunes, the operation, and traffic of vehicles are usually either impossible or accompanied by many problems and staggering costs. Crawlers and half-crawlers for power transmission instead of wheels make it possible to cope with the task in difficult transmission conditions. Table 3.5 presents the different types of crawlers and half-crawlers.

TABLE 3.5
Different Types of Crawlers and Technical Comparison

Type		Application	Specification	Advantages	Disadvantages
Rubber	Conventional	General purpose	Flexible, single-length track	Affordability, high comfort, low impact, high speed, high maneuverability	Low durability, low traction, sensitivity to temperature, no repairing
	Interchangeable	Hard-working purposes	Single unit-changeable		
	Interchangeable short pitch	Hard-working purposes	Easy to change and repair		
Steel	Link-and-Length	Applied around the running gear	Made out of styrene	Low wear and high durability, counterweight and balance, traction and load transfer, repair and maintenance, cleaning and upkeep	Equipment cost, surface damage, machine weight, noise factor, operator comfort
	Individual Links	General purpose	Greater balance and stability		
		Dragon's Smart Tracks	Easy to change and repair		

Previously, agricultural vehicles were equipped with wheels or crawlers. Some generations of agricultural vehicles, such as the EG series of Yanmar tractors, now have a combined transportation system called a half-crawler. This transportation system combines a wheel (in the front axle) and a crawler (in the rear), then the contact surface and the maneuverability increase. These moving units have a medium speed, a level of contact between wheels and crawler, and a moderate maintenance cost.

3.5.3 ROBOTIC FOOT

The robotic foot is one of the transportation systems which is not very common in AAVs. Asimo, HRP2 (Aist), Qrio, H7, Toyota Humanoid Robot, Hubo, and Wabian 2 are some robots that use robotic feet to move [101]. The main target of using feet for robots is interaction with human behavior, tasks, and duties and moving robots through complicated environments. These robots must be able to walk steadily in different environments. The legs are usually flat and stiff. An extra joint can have the flexibility to improve the robot's movement significantly. Adding more joints can improve the robot leg. There are different types of robot legs, which are compared in Table 3.6. The foot forms an element that ensures the interaction between the humanoid robot and the environment. Adding more joints increases flexibility and complexity but increases the application capability of the foot.

3.5.4 PROPELLER

Propellers are among the different transportation systems that fly AAVs in the air. Designing and manufacturing propeller blades are complicated and more redesigning, and optimization is needed to improve their performance. The modification makes changes to achieve maximum power, efficiency, and durability. Different materials have been used to make the propeller, such as wood, fiberglass, plastics, and metal. Wood became known as the first material for making aircraft propellers. Wood did not have much advantage to use, and all efforts were made to make it flexible. With the need to generate more power, wooden propellers became obsolete and were replaced by aluminum alloy propellers. These propellers had

TABLE 3.6
Different Types of Robotic Feet and Technical Comparison

Type	Complexity	Energy Use	Purpose
Plate foot	Low	Low	Single-purpose
Flexible foot	Medium	Medium	Multi-purpose
Active foot	Medium	High	Multi-purpose
Flexible active foot	High	High	Multi-purpose

TABLE 3.7
Different Types of Propellers and Technical Comparison

Type	Architecture	Complexity	Limitations of Specifications
Fixed Pitch Propeller	Angle (pitch) built	Medium	Low speeds, with limited range or altitude
Constant Speed Propellers	Variable pitch (angle)	High	It can be adjusted during flight
Ground-Adjustable Propellers	Manually altered pitch	Low	It cannot be adjusted when flying

high strength, low weight, easy repair, and the ability to rotate at high speed. With the advent of composite propellers, the weight of the propellers was reduced, and noise and vibration were reduced. The durability of this type of propeller was higher than other types. Table 3.7 presents three types of propellers and their specifications.

3.6 ACTUATING SYSTEMS AND OPERATION FUNCTIONS

All AAVs have actuating systems and controllable components (Figure 3.15). The number of components and controlling complexity in an AAV is much more than ordinary vehicles such as personal cars and buses. An AAV can have a robotic arm or end-effector as an actuating system for farm work such as harvesting, seeding, or watering. The controllable components of an AAV include steering control, forward and reverse movement, brake control, gear shift, rotational speed control, three-point hitch control, power axle, and external attachments. A vehicle's steering control or controllability is performed using the front axle rotation in all-wheel and half-crawler models. In crawler-type tractors and combine harvesters, the steering controlling apply using adjustment of rotation ratio. Until a couple of decades, complete vehicle control using controlling units (plus related circuits and sensors) in real roads and environments in some countries such as Japan and Germany was prohibited. Some regulations and policies remain. That was one of the reasons why full access to the controlling system of most commercialized vehicles is still unavailable. In a robot tractor, the required controllable control system is ordered from the company, and access will be available for laboratory use. Still, most countries do not allow these vehicles to cross the road.

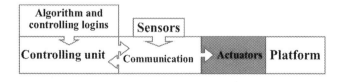

FIGURE 3.15 Actuator in autonomous agricultural vehicles.

FIGURE 3.16 Controllable components of a platform for robotic systems, (1) steering control, (2) forward and backward movement, (3) brakes, (4) gear changes, (5) rotational speed adjustment, (6) three-point hitch, (7) power take-off (PTO).

As the following operation function, controlling of power transmission ratio is essential to control the speed and torque of the vehicles. Gear change and moving forward and backward in older tractors and agricultural implements are manual. Electric controlling a manual power transmission system is almost impossible and not recommended for robotic applications. In newer models, there are automatic transmissions for shifting and motion direction (forward and backward), which is the best selection for intelligent controlling. There is no significant impact between controlling different types of automatic transmission. Different types of automatic power transmissions are the best selection for robotic applications. Figure 3.16 presents the controllable components of a platform for robotic systems.

Other operation functions are brakes (as restraint system), rotational speed (as engine torque and speed parameter), and three-point hitches (to connect other devices and equipment, plus empowerment of fixed devices). The main difference between tractors and other devices is that they have a three-point hitch and the power axis of tractors, which are not available in combine harvesters. All components must be electronically controllable using one or more ECU to develop an AAV. In some cases, the actuators and components can be controlled by adding a hydraulic/pneumatic cylinder or electronic motors such as a brake and hitch.

3.7 MOTION FUNCTIONS

Forward and backward motion, plus rotary speed, are three of the main operation parameters to move an AVV (Figure 3.17). Forward and backward motion apply the direction of AVV, and the rotary speed of the engine apply the velocity

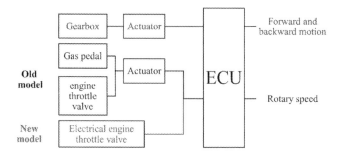

FIGURE 3.17 Motion functions of an AAV.

of motion. A gearbox directly controls the direction of motion. If the vehicle (commercialized) has an automatic gearbox, it is needed to find and control the related controlling function of the ECU. The communication between ECU and PC can be using RS232c. In the case of a prototype vehicle, the ECU is usually an open source or written by the designer. In the case of a commercialized vehicle, the company's permission to present related API and access is required.

About rotary speed, this parameter is directly controlled by the gas pedal. In the old version of vehicles, the gas pedal connects to the engine throttle valve using a wire. The engine throttle valve changes the air and fuel ratio based on the pedal displacement. To control rotary speed is needed to attach a linear actuator in the gas pedal or an angular actuator in the engine throttle valve. This job needs accurate calibration, repetitive tests, and high-level standardization. In a new development, the gas pedal is an angular sensor (potentiometer, encoder, etc.) that sends the angle data to an electrical engine throttle valve. The actuator of the engine throttle valve controls the air and fuel ratio using the data transmitted from the electrical gas pedal. This type of gas pedal is the best choice for an autonomous application. Because it can be easily controlled by the parameter of the electrical gas gate, this parameter is also controllable from the controlling program of ECU.

3.7.1 STEERING

The following parameter that controls the direction of AAVs is the steering angle. This angle is the angular change of steering in vehicles (in wheel-type transportation systems) and is the axis rotation difference (in crawler-type transportation systems). The steering system changes the vehicle's direction by applying rotational motion by the driver. The steering system must be able to operate with minimal force to be able to provide comfort to the driver. It must also be able to drive directly without causing any erratic movement. One of the essential factors in designing a steering system is the number of steering cycles compared to the amount of wheel rotation so that the angle transmission is more accurate. Also, the steering system should be able to absorb shocks and bumps as much as possible. Steering systems from mechanical steering to fully electric steering are currently used by various vehicles. The mechanical steering system is difficult for autonomous control because an electromotor with a robust gearbox is required. The installation of this system is complicated; it is not user-friendly, and also it is not suitable for commercial use.

In recent years, electric steering systems have been used in many vehicles, considering parameters related to comfort and safety. Introducing new technologies has resulted in energy savings, greater driver comfort, and reduced costs due to component depreciation. In the case of new designs, the electrical steering wheel has replaced the old systems. This type of steering system is electrically controllable and accurate. Table 3.8 presents the types of steering systems, including their specifications. Figure 3.18 presents a flowchart for the different types of electromechanical steering systems.

3.7.2 BRAKE

Stopping the vehicle when necessary is one of a vehicle's necessities and safety components. In AAVs, this task is performed by braking and its interaction with safety sensors. This component interacts directly with the safety range of the robot and commands the vehicle to stop if any obstacle enters. This function has to develop as simple as possible. Because the complexity of function, programming,

TABLE 3.8

Different Types of Steering and Technical Comparison

Type	Specification	Complexity	Limitations of Specifications	Controlling Ability
Mechanical steering	Mechanical components and joints	Low	Higher power requirements and lower welfare	Very low
Hydraulic Power Steering	Mechanical components and hydraulic components	Medium	Medium power requirement and medium durability	Low
Parameterizable Hydraulic Power Steering	Hydraulic components and hydraulic valves	High	Low power requirement for higher loads and requires high service and maintenance	Medium
Electrohydraulic Power Steering	Electrical components, hydraulic components, and electrical motors	High	Low power requirement, low energy usage, and higher welfare	High
Electro-mechanic Power Steering	Electrical components, mechanical components and joints, and electrical motors	High	Medium power requirement, medium energy usage, and higher welfare	High

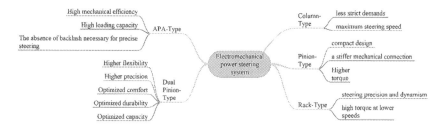

FIGURE 3.18 A flowchart for the different types of the electromechanical steering system.

equipment, etc., can cause a time delay, and each moment can cause a big disaster. Different types of brakes and their performance are presented in Table 3.9.

3.7.3 SHIFT

Shifting gears is essential in many vehicles, from bicycles to cars. Various shifting systems are available, each with a unique mechanism and different capabilities. Table 3.10 lists the types of shifting operators. Achieving a reliable system requires careful consideration of all its components, features, and benefits.

3.7.4 HITCHES

Some vehicles, such as tractors, have hitches to carry loads. The hitches are used to attach trailers, cultivation equipment, and stationary equipment. The connections available for AAVs can be important in portable loads and even equipment customization. Table 3.11 presents the types of trailer hitches and their characteristics.

3.7.5 PTO

A PTO is a mechanism that provides some actuating shafts to operate auxiliary equipment in certain vehicles. These shafts can receive their driving force from the engine using crankshaft timing gears. But in most systems, this power is usually taken from a part of the gearbox. Typical applications of PTOs include hydraulic pumps, compressors, generators, forklifts, cranes, rotating anchor wheels, fire hose pulleys, mixers, snow plow blades, and other mechanical mechanisms sourced separately. They need driving force.

The driving force of the PTO can be provided by one of the gears on the lay shaft, or the relevant shaft can be connected directly to the end of the lay shaft and rotate entirely from the shaft itself. Depending on the type of use, PTOs can be single speed or work at two speeds. A system such as an auxiliary gearbox is used in this case, which provides two different speeds. To achieve the desired shaft speed, the gear ratio, in this case, can also be selected as desired. Motion and energy conversion are critical in mechanics, which will happen in some car parts. PTOs have different types. Table 3.12 presents the different types of PTOs.

TABLE 3.9
Different Types of Brake and Technical Comparison

Type	Power Source	Complexity	Lifetime and Maintaining	Transmitted Power	Controlling Ability	Advantages
Hydraulic braking	Incompressible fluids	Higher than a mechanical system	Higher than a mechanical system with medium reliability	Higher than a mechanical system	Low	Higher durability
Electromagnetic braking	Electromagnets	Higher than other systems	Higher than other systems with a high reliability	Medium	High	Fast and cheap, and lower maintenance
Servo braking	Air intake system	Higher than a hydraulic system	Higher than a hydraulic system	High	Medium	High-power transmission and higher reliability
Mechanical braking	Mechanical components	low	High	Low	Low	Simple

TABLE 3.10
Different Types of Shifters and Technical Comparison

Type	Power Source	Complexity	Lifetime and Maintaining	Application	Controlling Ability	Advantages
Trigger Shifters	Manual	Low	High	Bicycle	Low	Simple
Manual Shifters	Manual	Low	High	Car, truck, and tractor	Low	Higher durability and simple
Twist-Grip Shifters	Manual	Low	High	Commuter and casual bikes	Low	Higher durability and ability to display the currently engaged gear
Downtube Shifters	Manual	Low	High	Bikes	Low	Simple
Bar-End Shifters	Manual	Medium	High	Bikes	Low	notably durable, simple, lightweight, and strong
Electrical Shifters	Electricity	High	High	Bikes, cars, and trucks	High	self-calibrating, eliminating rubbing issues, accurate, and high reliability

TABLE 3.11
Different Types of Hitches and Technical Comparison

Type	Complexity	Lifetime and Maintaining	Application	Controlling Ability	Advantages
Rear Receiver Hitch	Low	High	A common type of truck hitch	Low	Simple
Front Mount Hitch	Low	High	Cargo carrier, a winch, or a snow plow	Low	Similar and straightforward to rear receivers
5th Wheel Hitch	medium	High	Heavy-duty hitch	Low	Higher durability and heavy cargo
Gooseneck Hitch	Medium	High	Forward of the heavy rear duty	Low	Higher durability and heavy cargo
Pintle Hitch	High	High	Truck, and the lunette	medium	Heavyweights
Bumper Hitch	Low	High	Square receiver tube	Medium	Simple to use
Weight Distribution Hitch	Medium	Medium	Camping RVs	Low	Multitask

TABLE 3.12
Different Types of PTOs and Technical Comparison

Type	Complexity	Source	Application	Controlling Ability	Advantages
Transmission	High	Tractor transmission clutch	A common type of truck hitch	High	High performance
Overrunning clutch	Medium	Transmission system	Driven equipment	High	High performance
Live	High	Two-stage clutch	Multipurpose	Low	Higher flexibility in use
Independent	High	Separately controlling the PTO	Mechanical and hydraulic equipment	High	Higher flexibility in use with a single selector

3.8 POWER SOURCE

Internal and external combustion engines have long been a fascinating and challenging topic in energy conversion. Humans have always sought to purify the type of energy to use cheap energies, including wind, sunlight, water, sea waves, combustion from wood, coal, alcohol, petroleum products, etc. Available and easily usable energy, therefore, energy conversion units have been one of the essential needs of the engineering community. Meanwhile, internal and external combustion engines are among the equipment that can convert chemical energy into mechanical energy used in various industries in the past years. Figure 3.19 presents the main categories of combustion engines.

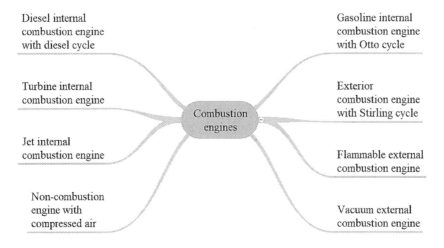

FIGURE 3.19 Different types of the combustion engines.

3.8.1 Vehicles with an Internal Combustion Engine (ICE)

Internal combustion engines can be categorized from different perspectives, such as the type of combustion, the arrangement, or the number of cylinders.

The first category classifies the types of internal combustion engines from their type of combustion. Based on this, there are two types of internal combustion engines: spark ignition and compression combustion. In each cycle of engine operation, the combustion process begins by using a spark, which is generated by a spark plug. In this way, a large voltage is established between the two electrodes of the candle. As a result of the electrical discharge between the two electrodes, it creates a spark. The combustion process begins with the compaction process. During the compression phase, the fuel-air mixture ignites spontaneously in the combustion chamber due to the high pressure and temperature of fuel and air caused by condensation.

The second classification for internal combustion engines is based on the thermodynamic cycle. Accordingly, internal combustion engines are divided into four categories: four-stroke and two-stroke. Four linear piston movements are required to perform each thermodynamic cycle in a four-stroke cycle: two reciprocating movements or two engine revolutions. Two linear piston movements are required to perform each thermodynamic cycle in a two-stroke cycle—a reciprocating motion of an engine rotation.

Classification of internal combustion engines based on the location of the valves are:

- Cylinder Head Valves—I Head Engine
- Valves in the engine block body—L-shaped motor or T-shaped motor
- One valve at the cylinder head (usually the inlet valve) and the other at the engine body—An F-shaped engine

Among all the types, the I-shaped model was the most common. And it can be said that it is a standard condition for today's passenger cars.

Generally, there are two main propulsion mechanisms for internal combustion engines. The reciprocating motion, in this case, is that the engine has one or more cylinders in which the pistons have reciprocating motion. The combustion chamber is at the end of each cylinder. In these engines, the reciprocating motion of the pistons becomes a rotational motion. And this kinematic conversion is done by a sliding crank mechanism. The rotational motion of the motor in this mechanism consists of a fixed body (stator), a non-centrifugal rotor, and an axis like the crankshaft. Each stage takes place in specific locations of the fixed body. So, with the help of the rotor, it forms a closed chamber in that place. The most famous rotary motor is the Vancle rotary motor. In internal combustion engines, the fuel-air mixture enters the cylinder chamber for combustion. The classification of internal combustion engines based on the air intake mechanism is as follows:

- Naturally Aspirated: In this engine type, there is no system to increase the inlet air pressure. And the inlet air is proportional to the cylinder's suction, and the piston enters the cylinder.
- Supercharged: In these engines, the inlet air pressure increases. This increase in air pressure is done by a compressor that draws its driving force from the crankshaft.
- Turbocharged: In this engine, the inlet air pressure increases as in the previous case. The difference is that the compressor turbine increases the air pressure driven by the engine exhaust gases.
- Crankcase Compressed: In this case, the two-stroke engines use the crankcase as the inlet air compressor.

3.8.2 Vehicles with Electric Motors

Electric vehicles (EVs) with an electric motor (EM) have been extensively researched as promising vehicles to reduce greenhouse effects. With advances in power, energy storage, and support, the plug-in hybrid electric vehicle (PHEV) can compete with the combustion engine vehicle in terms of driving and fuel consumption. Using optimal control strategies or the energy management system (EMS) concept, PHEV performance can be significantly improved. An EV is a vehicle that runs on an electric motor instead of an internal combustion engine that produces power by burning a combination of fuel and gas. Because of the high controllability of these vehicles, their robotization was desired. The electric vehicle can be classified into four different types as Table 3.13, considering the power and propulsion.

TABLE 3.13
Classification of Electrical Vehicles Considering Power and Propulsion

Type	Sub-Classification	Source of Energy	Power	Emission	Controlling
BEV	–	Electricity	Electromotor	Almost zero	Simple
HEV	Series	Fuel, Electricity	Electromotor + engine	Medium	Medium
	Parallel	Fuel, Electricity	Electromotor + engine	Medium	Medium
	Series- Parallel	Fuel, Electricity	Electromotor + engine	Medium	complicated
	Complex	Fuel, Electricity	Electromotor + engine	Medium	complicated
FCEV		Hydrogen	Electromotor	Low	Simple
ICEV	Diesel	Fuel (diesel)	ICE	High	Extremely complicated
	Gasoline	Fuel (gasoline)	ICE	High	Extremely complicated

3.8.2.1   Battery Electric Vehicle (BEV)

Electric vehicles using BEV technology run entirely on a battery-powered electric propulsion system known as all-electric vehicles. The electricity used to move the vehicle is stored in a large battery pack that can be charged by connecting to the mains, and when the battery is charged, one or more electric motors are powered to start the electric car. The main components of BEVs are an electric motor(s), inverter, battery, control module, and propulsion. The principle of BEV's work is that the electric motor's power is converted from a DC battery to AC. By pressing the accelerator pedal, a signal is sent to the controller. The controller adjusts the vehicle speed by changing the AC power frequency from the inverter to the motor. The engine then rotates the wheels through a gear. If the brake pedal is depressed or the electric vehicle slows down, the engine turns into an alternator and produces power that is returned to the battery.

3.8.2.2   Hybrid Electric Vehicle (HEV)

Hybrid or hybrid electric vehicles (HEVs) work with an electric motor and an ICE. In other words, cars and tools have more than one source of movement. HEVs have more mileage than BEVs due to having an ICE to charge the batteries. HEVs are divided into four groups based on drivetrain setup: series, parallel, series-parallel, and complex hybrids. The main components of HEV are an internal combustion engine, electric motor, battery with controller and inverter, fuel tank, and control module. The principle of HEV is that the internal combustion engine is fueled by gasoline, and a battery powers the electric motor. The internal combustion engine is used for both driving and charging discharged batteries. The wheels are driven by a gearbox that spins the internal combustion engine and the electric motor at the same time.

In a series of hybrids, an electric motor moves the vehicle, and a small-sized ICE charges the batteries. These HEVs are ICE-assisted electric vehicles. In parallel hybrids, the vehicle moves using ICE and/or an electric motor, alone or by both combined. A power-split hybrid can change between parallel or series modes and use the advantage of each. In complex HEV, two bidirectional power converters are utilized, one for the main electric motor and another for the auxiliary electric motor. Unlike in series-parallel HEVs, both of these motors can propel the wheels concomitantly.

3.8.2.3   Plug-in Hybrid Electric Vehicle (PHEV)

These types of electric vehicles are also known as series hybrids, which use an internal combustion engine and a rechargeable battery with an external socket (they have two plugs). It means that mains electricity can charge the car battery instead of the motor. Combustion engine fuels include conventional fuels (such as gasoline) or alternative fuels (such as biodiesel). There are two modes for PHEV vehicles: (1) All-electric mode (PEV), in which the engine and battery supply all the energy of the car, and (2) hybrid mode (PHEV), in which both electricity and gasoline/diesel are used. The main components of a PHEV are an electric motor, internal combustion engine, inverter, battery, fuel tank, control module, and battery charger.

The working principle of PHEVs is that they run in all-electric mode and use electricity until their battery runs out. When the battery is discharged, the engine takes over, and the car acts as a standard, non-plug-in hybrid. PHEVs can be charged by connecting to an external power supply, using a motor or regenerative brakes. When the brake is applied, the electric motor acts as a generator and uses energy to charge the battery. An electric motor complements engine power. As a result, smaller engines can increase vehicle fuel efficiency without compromising performance.

3.8.2.4   Fuel Cell Electric Vehicle (FCEV)

This electric vehicle uses "fuel cell technology" to generate electricity. The main components of FCEV are an electric motor, fuel cell, hydrogen storage tank, battery with converter, and controller. The principle of FCEV is that the fuel cell is the primary energy source, and the critical technology for FCEV is an electrochemical device that generates DC electrical energy through a chemical reaction. There are five main components in a fuel cell: the anode, anode layer, electrolyte, cathode, and the cathode catalyst layer. Connecting the appropriate parallel/series of fuel cell sources can generate the force required to drive the vehicle.

3.9   POWER TRANSMISSION

Modern cars with internal combustion engines had several fixed parts from the earliest stages of construction in 1879. One of these parts was the transmission system. Power transmission is required to move the car. This system is essential in cars and is one of the most sensitive technical parts. Its repairs are costly and troublesome, and its maintenance requires excellent care. Repairing it is a very specialized job; not everyone can afford it. The transmission system has changed a lot during the history of the automotive industry and is one of the most important technical parts of the automobile. When introducing and expertizing cars made by car companies, their transmission system is usually examined after the engine, and their type and characteristics are discussed, which shows the importance of this system in the car.

Today, the transmission system has different types used according to the car's performance. In simple terms, the car's power transmission system has the vital task of transferring the power generated by the engine to the wheels to rotate them and thus move the car. It transmits the mechanical force generated by the car's propulsion to the wheels. Therefore, this system's various parts and components together establish the connection between the motor and the wheels. In other words, any part between the engine and the wheels is part of the transmission system. Switching on and off, changing engine power, and adjusting wheel speed are also essential tasks of the transmission system. The importance of the car's power transmission system increases because the power transmission and engine production power must reach the wheels with a minor reduction and waste. Also, the transmission system controls the car's speed and acceleration simultaneously as the engine power transmission. But this system also has other tasks, which are:

- Power off: This system must also be able to turn off the power when necessary.
- Engine power change: The transmission system must be able to change the engine power output (power and torque) and adjust to different driving conditions. Then transfer this power to the wheels.
- Adjusting wheel speed: Wheels should also have different speeds depending on road conditions and in terms of turns, roughness, and friction. The task of adjusting these cycles based on the output power of the engine is also with the power transmission system.

Each task performed one component of the gearbox and the transmission of power in the car entirely and flawlessly. Therefore, preventing depreciation and damage is also of high importance in the list of ordinary car care.

The components in the transmission system include a clutch system, gearbox, auxiliary gearbox (used only for four-wheel drive vehicles), steering shaft (only for rear-wheel drive vehicles and four-wheel drive), differential, pluses, and drive wheels.

The power transmission system is divided into four categories:

- Front Wheel Drive (FWD): In this type of transmission, the car engine's power is transferred to the two front wheels. In this system, the gearbox, steering wheel, and transmission axles are located in the front of the car. The cabin space is less occupied. The engine is placed vertically inside the car. Cars made with this system are called front differentials.
- Rear Wheel Drive (RWD): In the rear differential system, the energy generated by the steering rod coming out of the gearbox is transferred to the car's rear wheels. The engine positioning system in rear differential vehicles is such that the engine is positioned parallel to the vehicle's longitudinal axis, and the steering wheel is connected directly to the rear differential.
- Four-Wheel Drive (4WD): The meaning of this phrase stands for four wheels move. This system is the oldest four-wheel drive transmission system. Engine power is transferred to the car's four wheels by using two gearboxes and two differentials, and they move the powerful off-road vehicles.
- All-Wheel Drive (AWD): As the name implies, in this system, the power transmission is done simultaneously to the four wheels of the car, and all the wheels of the car move. The performance of this system is the same as the performance of the 4WD system. The difference is that in this system, three electronic differentials are used to control the force applied to the vehicle axles, and the differential of each axle is responsible for managing the transmission power between the wheels.

This system is commonly used in trucks and urban and sports crossovers because using this system improves the steering and balance of the car. It is one of the reasons why the experience of driving a crossover car is such an enjoyable experience. This transmission system operates in such a way that it distributes energy evenly between the front and rear wheels. Of course, this process takes place at low speeds, and if the speed increases, all the power is transferred to the car's front wheels, because at high speeds, any misplaced movement of the rear wheels can cause an imbalance in the car.

NOTE

1 Research and Development.

4 Sensors

4.1 INTRODUCTION

It is inconceivable for a human being to be senseless. Think of it as having no sense of sight, no sense of taste, or even a sense of touch, smell, or hearing. Our communication with our surroundings is with these five senses. AAVs are no exception to this rule. They have to see the environment to be able to react to the different conditions that prevail around them. They have to interact with environments and other robots. They must protect their safety zone and not damage others. They need to know where they are (position), where are going (target and direction), and what path need to take to reach their goal (route). Sensors as interfaces of AAVs with the environment can convert the required parameters into signals (digital or analog) for their processor/controlling unit. So with this introduction, it can be claimed that sensors are one of the essential parts of the AAV. Figure 4.1 presents sensors in the configuration of autonomous agricultural vehicles.

Different classifications for sensors can be provided based on energy contribution (modulator or generator), the output signal (analogor digital), or operation mode (defection or comparison). Also, we can classify the sensors based on measurand (acoustic, electrical, optical, biological, chemical, radiation, thermal, magnetic, or mechanical), energy (active or passive), signal conversion (chemical, biological, or physical), physical (contact, or non-contact), comparability (absolute, or relative), or material. As illustrated in Figure 4.2, the required sensors in AAVs are divided into three main groups in this book, which is more applicable in the esign of AAVs: (1) positioning sensors to indicate the exact location/position/coordination, (2) attitude sensors to indicate the direction/ orientation and inclination, and (3) safety sensors for detecting and protecting the safety zone.

4.2 POSITIONING SENSORS

Positioning sensors can track an object's motion or calculate its relative position compared to a predetermined reference point. These kinds of sensors can also determine whether a thing is there. It is important to note that other sensor types have functions comparable to those of position sensors. Positioning sensors can take action when they detect an object moving (such as illuminating a floodlight or activating a security camera). The presence of an object inside the sensor's range can also be detected using proximity sensors. As a result, both sensors may be categorized as specialized position sensors. Our associated guides on proximity and motion light sensors include further information. Positioning sensors

DOI: 10.1201/9781003296898-4

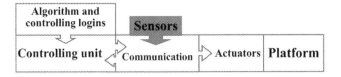

FIGURE 4.1   Sensor in the autonomous agricultural vehicles.

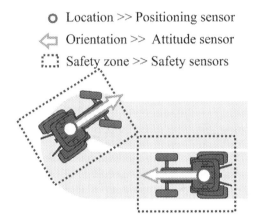

FIGURE 4.2   Sensors required for farming robots: positioning, orientation, and safety sensors.

differ in that they focus primarily on identifying an object and logging its position. As a result, they use a feedback signal containing location information. Different positioning sensor types exist. The many kinds of positioning sensors are categorized in Table 4.1. The many positioning sensor applications are shown in Figure 4.3.

But in this book and all AAVs, the positioning sensors provide the location parameters on the ground, such as time, longitude, latitude, and altitude. This sensor is one of the most essential and central sensors for autonomous navigation. In recent decades, the global navigation system has been added to our everyday tools, gadgets, and even vehicles so that we can navigate with it. It should be noted that this technology was used exclusively for military applications until the 1990s, and even later, to use it, you needed manual GPS. This process was a standard path for the development of technology.

Although for daily life, such as navigation, the accuracy of the satellite navigation system is not given much attention in agricultural applications, this issue has been and is essential. An accuracy of 50 cm may be enough for navigating urban and interurban roads, but this accuracy is a disaster for agricultural applications. In the worst case, this accuracy should not exceed 20 mm. This parameter led researchers to look for ways to develop navigation systems and increase their accuracy. In the following, we will discuss different positioning

TABLE 4.1
The Different Types of General Positioning Sensors

Type	Variety	Mechanism	Pros and Cons
Potentiometer	Linear or rotary	Elastomer damped wipers and a resistive track	Low accuracy and limited bandwidth
Cable Extension Transducer	Linear position	Cable and a spring-loaded spool	Compact, suitable for limited spaces, and measures large ranges of position
Hall Effect	Non-contacting position sensor	Magnetic fields	High accuracy and wear-free
Eddy Current	–	Magnetic fields	Well suited to dirty and noisy environments and works in high temperatures
LVDT and RVDT	Non-contacting sensors	Metal core and coiled wires	Long life and high accuracy
Positek PIPS	Non-contacting sensors	Fewer coils in comparison with LVDT and RVDT	Better stroke-to-length ratio, high accuracy and are suitable for intrinsically safe environments
Laser Position Sensors	Non-contacting sensors	The sensor emits a laser	A wide range of applications
Capacitive Sensors	Non-contacting sensors	Two parallel conductive plates with a voltage variation	No chance of wear, high accuracy, and wide range of application

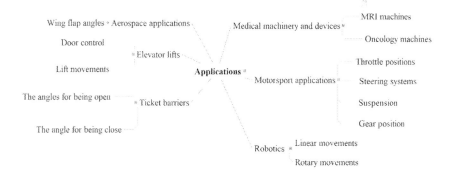

FIGURE 4.3 The applications of positioning sensors.

systems for agricultural applications and their development procedure. In this case, many positioning systems such as video sensors, lateral navigation (LNAV), smooth navigation (SNAV), X-ray pulsar-based navigation (XNAV), a global positioning system (GPS), differential GPS (DGPS), and real-time kinematic GPS (RTK-GPS) will be discussed. Each positioning sensor has a different methodology, accuracy, time delay, and price. Some of them are still used, and the advancement of technology does not justify others.

4.2.1 NAVIGATION USING A VIDEO SENSOR

The first attempts by researchers to develop a positioning system to control AAVs were made by Ishii et al. [102]. As you can see in Figure 4.4, the structure of this system consists of two fixed video sensors installed laterally at a distance (L), lights installed on a mobile platform, PSs, and wireless modems installed on the mobile platform and station.

The system operates using a triangulation rule which is a method in trigonometry and geometry. It calculates the coordination of a point by measuring the angle of one point to two. This method is used for three-dimensional optical measurements. In this system, two cameras (as two points) are fixed relative to each other to capture the desired point, the mobile vehicle. The cameras follow the points on the vehicle and extract the lateral rotation angle using optical sensors. Two red and white light fluorescent indicators are installed on the robot, and the robot's position is detected by rotating the optical sensors. Optical sensors consist of a rotary switch, a CCD camera, and a stepper motor connected to a computer using the RS232c connection method. The control system sends the position to the robot's PC for analyzing the radio receiver and a wireless modem. The location of the vehicle is calculated using triangulation, which will be explained using the following equations (Figure 4.4).

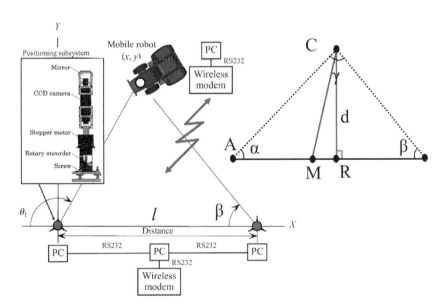

FIGURE 4.4 Navigation using video sensor (left: structure and mechanism, and right: triangulation rule used in operation).

As illustrated in Figure 4.4, there is a rule in the triangle method:

$$l = \frac{d}{\tan \alpha} + \frac{d}{\tan \beta} \rightarrow d = \frac{l}{\left(\frac{1}{\tan \alpha} + \frac{1}{\tan \beta} \right)} \qquad (4.1)$$

We can also write using the law of sines:

$$\frac{\sin \beta}{AC} = \frac{\sin \gamma}{l} \rightarrow AC = l \cdot \frac{\sin \beta}{\sin \gamma} \qquad (4.2)$$

And also:

$$d = AC \cdot \sin \alpha \qquad (4.3)$$

And size d is calculated as follows:

$$d = l \cdot \frac{\sin \beta \cdot \sin \alpha}{\sin \gamma} \qquad (4.4)$$

The following formula is used to measure the MR distance:

$$MR = BM - RB = \frac{l}{2} - d \cdot \cot \beta \qquad (4.5)$$

4.2.2 LATERAL NAVIGATION (LNAV)

The Kubota Company presented lateral navigation (LNAV) or transverse navigation. This naming was because positioning was done laterally (Kondo, Monto, et al., 2011). The LNAV could detect the position of a device based on wires stretched around the ground and electromagnetism generated by the electric current of the wires (Figure 4.5). The impact of the magnet was related to the distance from the wires. This rule provided the basis for locating the device or using LNAV. This system has two magnetic fields on either side of the device. It was necessary to train the robot manually in different places inside the magnetic field. The advantage of the system was its all-weather usability. After a while, it was stopped in use due to the costly implementation on large farms and low accuracy. Although the creativity in positioning this system was not very practical for the defined application, this technique can be used in similar cases in low areas.

4.2.3 X-RAY PULSAR-BASED NAVIGATION (XNAV)

The X-ray pulsar-based navigation (XNAV) system designed by BRAIN-IAM and developed by Sanyo Electric (Figure 4.6). This system was more like video

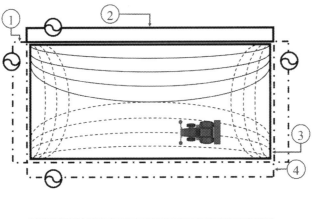

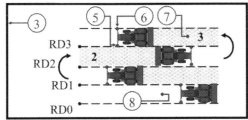

FIGURE 4.5 LNAV positioning system, (1) Cable C to detect end-to-end and ambient (9.8 kHz), (2) Cable A to return and work with ambient (4 kHz), (3) field boundary, (4) Cable B To return and work with the environment (1.5 kHz), (5) left sensor, (6) suitable sensor, (7) unmanned return operation, (8) training implementation.

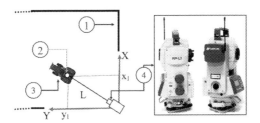

FIGURE 4.6 XNAV positioning system, (1) field boundary, (2) target, (3) mobile station, (4) Topcon AP-L1 reference station.

sensors, but these navigation systems used an optical measurement system [103]. Reference stations observe the target embedded in the mobile device, and the diagonal distance (L) and horizontal angle are measured using sensors embedded in the reference station. The target's position (mobile device) can be calculated using geometric formulas.

4.2.4 GLOBAL POSITIONING SYSTEM (GPS)

The global positioning system (GPS) is a satellite-based radio navigation system developed by the U.S. government and operated by the U.S. Space Force [104]. In the early 1970s, U.S. Department of Defense experts introduced GPS projects. This project met the U.S. government's requirements in positioning anywhere on the planet at any time and in any weather conditions with high accuracy. GPS stands for global positioning system and is a satellite positioning system that allows us to determine the position in all parts of the globe in terms of latitude and longitude. Initially, the plan was launched by the United States for military activities and was gradually made available to the public for various applications, including mountaineering, travel, and positioning needs. The U.S. DoD's (Department of Defense) satellites in orbit serve as the foundation for the navigation system. Although this technology was primarily created for military use, it was made available for civilian usage in 1980 by the government. This technology operates 24/7/365 in any setting, anywhere around the globe. The 24 satellites that make up the GPS rotate once every 12 hours worldwide to provide global time, location, and velocity data. GPS's primary function is to precisely identify the areas on the globe by determining the distance from the satellites. This system lets you create or otherwise record exact locations in the world and assists you in navigating from those locations. This system was mainly designed for military applications, but in 1980, it was made accessible for civilian use.

GPS is a satellite system based on 24 satellites (21 main satellites and three spare operational satellites) operating in six orbits at an altitude of 20,200 km above the ground. To be able to determine the exact position on the ground. A GPS receiver receives a signal from a group of satellites, calculates its position, and shows the geographical coordinates of the desired location on the screen. These coordinates indicate where the target is on the map. When using GPS, the accuracy of the location of each point is important. The optimum accuracy for a climber or a soldier in the desert is 15 m. For a ship near the waterfront, the proper accuracy is 5 m. For a ground surveying engineer, the optimum accuracy is 1 cm. For agriculture applications, the desired accuracy is 2 cm. All the accuracy mentioned above can be achieved using different GPS systems. Only the difference between the GPS receiver and the type of technique is used.

GPS uses a method called "triangulation" or "triple calibration" and requires a message from at least three satellites, preferably four satellites, to accurately detect the position. A GPS receiver measures the distance between itself and each satellite to triangulate. Atomic clocks are used on each satellite to measure transmission time. Along with the distance, the device needs to know exactly where the satellites are at any given time.

In general, a GPS consists of three segments: (1) space segment, (2) control segment, and (3) user segment. The space segment is the constellation's number of satellites. It consists of 29 satellites orbiting the globe at 12,000 miles every

12 hours. The route/navigation signals are sent by the space segment, which also stores and retransmits the control segment's route/navigation message. Atomic clocks on the satellites, which are highly reliable, regulate these transmissions. The GPS space segment comprises a satellite constellation with enough satellites to guarantee that users can always see at least four satellites at once from any location on the planet's surface.

The control segment comprises a control center and five monitoring units, each with an atomic clock dispersed worldwide. The master control station receives irregularities and updates them before relaying the corrected signals to the GPS satellites through ground antennas. The five monitor stations continuously monitor the GPS satellite signals. The control section is additionally known as a monitor station.

The GPS receiver, which receives signals from GPS satellites and measures its distance from each one, is part of the user segment. This sector is primarily used by the U.S. military, missile guidance systems, and civilian uses of GPS in practically every industry. Most civilians use this for surveys, transportation, and management of natural resources, followed by uses in agriculture and mapping. This information is stored inside the GPS receiver. The advantages and disadvantages of GPS are shown in Table 4.2.

4.2.5 SMOOTH NAVIGATION (SNAV)

Other device positioning systems include SNAV or Psycho Positioning, developed by the Japanese aviation electronics industry. Figure 4.7 shows that the system benefited from differential GPS, TMS, and the IMU for navigation [103]. TMS and IMU sensors are used to correct system interruptions and positioning errors. The accuracy of this system was good, but its price and support services were expensive. Also, it depended on the reference station among the obstacles to using these systems.

TABLE 4.2
Advantages and Disadvantages of GPS

Advantages	Disadvantages
It is dangerous to adjust it while driving.	With the right equipment, its accuracy can be increased up to 30 cm.
Sometimes it can lead to bad roads.	
It may not recognize your destination.	There is no need for a person to use a paper map.
Tall buildings or forests can block the signal.	It can offer different routes, such as the fastest,
If satellite navigation stops working, you need to be able to read the map (and a paper map).	maximum highway speed, scenic route, etc.
Storms and bad weather can affect accuracy.	Saves time by identifying privacy.
Maps need to be updated to keep track of road changes.	It can have a map of any country.
	A destination can be saved without the need to reset.

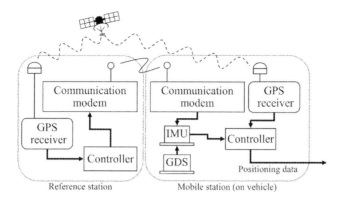

FIGURE 4.7   SNAV positioning system.

4.2.6   DIFFERENTIAL GPS (DGPS)

Civilian users had to deal with selective availability (SA) until 2000. To reduce the usefulness of GPS and the potential for U.S. adversaries to abuse it, the DoD purposefully added random timing flaws to satellite signals. The accuracy of measurements could be affected by these timing mistakes by as much as 100 m. A single GPS receiver from any manufacturer can reach an accuracy of about 10 m when SA is removed. Differential correction of the data is necessary to attain the accuracy levels required for high-quality GIS records, which range from one to two meters to a few millimeters. Most GPS-gathered data for GIS is differentially adjusted to increase accuracy.

The fundamental tenet of DGPS is that atmospheric errors will be comparable for any two receivers that are substantially close to one another. A GPS receiver must be installed at a precisely known position to use DGPS. The base station or reference station is this GPS receiver. Based on satellite signals, the base station receiver determines its position and compares it to a known location. The second GPS receiver, the wandering receiver, records GPS data and applies the difference. Using radio signals, the corrected information can be applied to data from the roaming receiver in real-time or after data gathering using specialized processing software.

DGPS uses a well-known fixed position to adjust GPS signals in real time to eliminate pseudorange errors. Pseudorange errors are caused by signal delays in the passage of atmospheric layers. GPS signals coming down from satellites to Earth must pass through the Earth's atmosphere, so they are delayed and affect the time it takes to transmit a signal from each satellite to the GPS receiver. This delay causes a minor error in the GPS engine and causes an error in the measured position.

GPS signals must pass through the outer layer of the Earth's ionosphere. This layer of the atmosphere is separated from the particles by solar radiation, has a positive charge, and has the most significant effect on the passage of electromagnetic signals, such as radio signals. The ionosphere layer adds a relatively significant delay. Some factors estimated, allowed, and related to the delay of up to 16 nanoseconds for the transmission signal. Of course, this is a variable that is variable and can cause up to 5 m of error in the position taken.

The second layer through which GPS signals pass is the troposphere. This part of the atmosphere includes clouds, rain, and lightning. It adds a much smaller delay (1.5 nanoseconds), which can have a position error of up to half a meter. These are random delays that fluctuate, and there is no way to measure their amount at any given time accurately. The latency of each satellite can also vary depending on the weather conditions of the regions.

A static base station can provide signal delay correction messages by setting the base station to a specific point on the ground and then determining its exact position. It is done by recording GPS information for as long as possible. Over time, information from the base station changes the ionosphere and troposphere, causing signal latency to change randomly. Because these delays are subject to random changes, they can be averaged.

Another form of signal correction is the use of fixed satellite signals. These satellites follow the Earth's rotation direction a certain distance above the Earth's equator. Objects in Earth's orbit remain stationary in the sky when viewed from Earth. A network of ground-based stations sends data to a central computer that calculates errors in the current location of the area. This information is sent to a fixed Earth satellite before replaying to GPS receivers. The correct correction can be calculated based on the position of the receiver. Although it does not increase the positioning accuracy as much as using a fixed base station that performs a long record of information, it does not require any adjustments, so it is quick and easy to use anywhere.

The general term for these systems is SBAS, which stands for space based augmentation system for different countries as follows:

- United States: WAAS.
- Europe: EGNOS (European Geostationary Navigation Overlay Service).
- Japan: MSAS (Multi-Purpose Satellite Amplification System).
- India: GAGAN (Geo-enhanced Navigation System).

4.2.7   REAL-TIME KINEMATIC-GPS (RTK-GPS)

Current satellite navigation (GNSS) systems use real-time kinematic positioning (RTK), an application of surveying, to adjust for frequent inaccuracies. In addition to measuring the signal's information content, it also measures its carrier wave phase. It relies on a single reference station or an interpolated virtual station to offer real-time adjustments and can achieve centimeter-level accuracy (see DGPS) [105]. The technology is frequently referred to as carrier-phase enhancement, or CPGPS, concerning GPS [106]. It can be used for unmanned aerial vehicle navigation, hydrographic, and land surveys. The amount of time it takes for a signal to travel from a satellite to a satellite navigation receiver can be used to determine how far apart they are. To determine the delay, the receiver must line up a pseudorandom binary sequence in the signal with a pseudorandom binary sequence it has internally created. The satellite's sequence is behind the receiver's sequence because it takes some time for the satellite signal to get to the receiver. The two sequences are eventually synchronized by delaying the receiver's sequence progressively. The accuracy of the resulting range measurement depends on the receiver's electronics' ability to decode satellite signals correctly and on other error causes, such as non-mitigated ionospheric and tropospheric delays, multipath, satellite clock, and ephemeris errors [107].

In reality, RTK systems use a few mobile devices and a single base-station receiver. Mobile units contrast their phase measurements with the information received from the base station as it rebroadcasts the carrier phase it has observed. A correction signal can be transmitted from the base station to the mobile station. The most common real-time and inexpensive transmission method is to utilize a radio modem, usually in the UHF band. In the majority of nations, particular frequencies are allotted for RTK usage. A built-in UHF-band radio modem is typically available in field survey equipment. Up to 20 km from the base station, RTK improves accuracy [108].

By the time GPS/GNSS signals reach a receiver, imperfections are introduced by various factors, including errors in satellite clocks, inaccurate orbits, travel through the layers of the atmosphere, and many more. With the significant exceptions of multipath and receiver noise, real-time location is based on the assumption that these GPS/GNSS error sources are connected. Nevertheless, since the faults are unpredictable, the only approach to fix them is to keep an eye on them as they occur. Installing a GPS/GNSS receiver on a base station whose precise location is known is an excellent approach. The computer in this base station receiver can determine its position from satellite data, compare it to its absolute known position, and determine the discrepancy.

Over a data link, the base and rover can exchange the resulting mistake corrections. It works well as long as the base station keeps an eye on them constantly, or at least while the rover receiver or receivers are operational. While this is going on, the rovers travel around collecting the data from the sites whose locations you want to know concerning the base station, ultimately, the goal.

GPS, DGPS, and RTK-GPS are generally based on global navigation satellite (GNSS) technology, which differs in positioning, accuracy, topography, and application. The GNSS system is available in the United States, Japan, Russia, India, China, and Europe with NAVSTAR, QZSS, GLONASS, IRNSS, BeiDou-3, and GALELEO systems, respectively.

As shown in Figure 4.8, the RTK-GPS, or real-time kinematic-GPS system, increases the accuracy of GPS signals using a fixed base station that transmits corrections wirelessly to a mobile receiver. The GPS motor can adjust the antenna's position by 1 to 2 cm. This method involves measuring the phase of the satellite signal carrier, which is then subjected to complex statistical methods to balance these signals to eliminate most of the typical GPS-type errors. This alignment process has three stages: "floating," ambiguity, and "fixed" ambiguity.

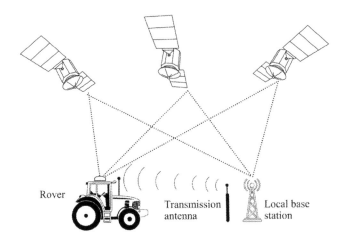

FIGURE 4.8   Implementation and operation of RTK-GPS.

The accuracy in floating and stationary modes varies between 0.2–0.75 m and 0.02–0.01 m, respectively. The correction signal is sent at 1-second intervals. The mobile stand or system attached to the moving device is mounted on a moving vehicle, in which case corrections need to be sent quickly to the GPS receiver to reduce accuracy to about 3–5 cm.

The GPS requires three satellites that provide latitude and longitude with an accuracy of about 3 meters. The mobile device or robot carries a GPS receiver, antenna, or mobile phone. In the DGPS system, the device's position is corrected using a fixed positioning station, and its accuracy is increased up to 10 cm. Specifically for agricultural work, a minimum accuracy of 5 cm is required. This requirement has only made the only application system in this field RTK-GPS (Figure 4.8). The mobile device or robot must carry a GPS receiver and a radio antenna in this system. Companies now offer both systems in one package.

4.3   ORIENTATION SENSORS

Orientation or attitude sensors are an integral part of the automatic control in an AAV and complement the main structure of a robot's automatic control. Although a robot needs to use positioning sensors to detect its position in a two-dimensional environment, its direction is one of the parameters determined by the orientation sensors. These sensors not only help to orient the robot but also with different algorithms and, with the support of sensor combination and filters (such as the Kalman filter), can detect the errors of positioning sensor signals into tunnels or nearby buildings, trees, and obstacles and then correct it.

The primary purpose of using orientation sensors is to measure the central angles in three dimensions (θ_{Yaw}, θ_{Roll}, θ_{Pitch}), angular velocities in three dimensions ($\dot{\theta}_{Yaw}$, $\dot{\theta}_{Roll}$, $\dot{\theta}_{Pitch}$), and angular acceleration in three dimensions ($\ddot{\theta}_{Yaw}$, $\ddot{\theta}_{Roll}$, $\ddot{\theta}_{Pitch}$) (Figure 4.9). The yaw angle is mainly used to control robots

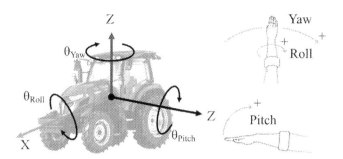

FIGURE 4.9 AAVs principal axes.

and sometimes pitch and roll; angular velocity measurements and angular accelerations can help formulate algorithms with higher accuracy.

In the early days of AAVs, geomagnetic direction sensors (GDSs) were used to indicate deflection and yaw. These sensors had two main problems: their accuracy changed under the influence of the surrounding magnetic field, and the angular deflection of the device when passing obstacles and the back of the heights caused an error in measuring the angles. That's why fiber optic gyroscopes (FOGs) became popular. Although these sensors were much more accurate than GDSs, they were not economically viable for agricultural use due to their high cost and used in aircraft and ship navigation. Finally, the construction of inertial measurement units (IMUs) has prompted researchers to replace these orientation sensors with previous technologies due to their reasonable price, good accuracy, and greater angle measurement. Due to their ability to measure quadrants, directions in three dimensions, linear accelerations, and the Earth's gravitational force, these sensors showed high capability and efficiency in conducting research and operational projects. These sensors consist of a gyroscope and an accelerometer; in some models, a magnetometer has been used. Precisely combining a good IMU with a RTK-GPS and a custom sensor combination technique controls various devices with great precision. It has led to the recent production of RTK-GPS companies to include an IMU in the package. The development of these sensors is happening very fast. More innovative systems may have been produced as you read this book. In the following, we will examine each of the orientation sensors in more detail.

4.3.1 GEOMAGNETIC DIRECTION SENSORS (GDSS)

Since the Earth's magnetic field is also known as a geomagnetic field, the direction can be determined by measuring the Earth's magnetic field. Geomagnetic sensors detect the Earth's magnetic field and can be used as an electronic compass for various purposes. Geomagnetic sensors can be two-axis (X and Y) or three-axis (Z, Y, and X). There is no need to consider the slope when measuring the direction with geomagnetic sensors to a simple electronic compass, so only the X and Y axes are used. However, to solve the angular and tilt problems of the sensor, it is necessary to correct the three axis values of the geomagnetic sensor with an accelerometer to correct it in the right direction. One of the significant problems with this type of sensor is that it fails in the presence of other magnetic fields, such as transformers, power plants, transmission lines, vibrating systems, car engines, and any device that generates magnetic waves. Using these sensors in automated systems has made them obsolete.

A magnetic sensing element could be a sensing element capable of measuring the magnitude and direction of a field. A variety of sensors exist reckoning on the aim; however, the subsequent square measure is a typical example. One is the Hall sensor. A Hall sensor is a sensor that uses the Hall effect to measure the magnetic flux density and produces a voltage proportional to the magnetic flux density. It is simple and frequently used in contactless switch applications like door open/close sensing and laptop computers. The next one is the magneto-resistive (MR) sensor. An MR sensor determines the intensity of the field through adjustments in individual electrical resistors dependent on the magnetic field. Due to its greater sensitivity and lower power consumption than Hall sensors, this magnetic sensor is utilized rather frequently. MR sensors are employed in applications such as motor rotation and location detection and electronic compasses' geomagnetic detection. The next-generation magnetic sensor, known as a MI (magneto-impedance) sensor, uses the magneto-impedance properties of a unique amorphous wire. It has more than 10,000 times greater sensitivity than Hall sensors, enabling precise monitoring of even the slightest changes in the geomagnetic field. Applications requiring great sensitivity, such as metal foreign object identification, indoor location, and ultra-low-current azimuth detection, are well suited (e-compass).

4.3.2 Fiber-Optic Gyroscopes (FOGs)

Based on the Sagnac impact, FOGs are passive systems that use light to calculate movement. In opposite guidelines, rotation movements are measured by sending identical laser beams through an extended fiber-optic coil. The laser beam visiting the path of rotation reviews a slightly shorter path delay than the alternative beam, measuring the segment shift among the two beams' well-known modifications in orientation. Gadgets are loose to rotate in three dimensions, iXblue's Fiber-optic-based inertial navigation systems carry three FOG gyroscopes, measuring rotations on the three exceptional axes [110,111].

FOGs, as their name implies, measure the rotation of a moving device. In the past, mechanical gyroscopes were used to measure the device's rotation. Rotary equipment is less used. Today, laser gyroscopes (RLGs) and fiber-optic gyroscopes (FOGs) are a priority. However, laser gyroscopes have been practically rejected due to the problems in preparing the raw material of the active laser medium and the high cost of accessing its production line. A fiber-optic gyroscope is the most used in guided systems because of its fast response, low volume and weight, lack of rotating part, relatively good bias stability against mechanical gyroscope, and not being affected by electromagnetic noise due to the use of insulated optical fiber.

The fiber-optic gyroscope, or FOG, senses a change of orientation using the Sagnac effect and performs the mechanical gyroscope function. But its principle operation is based on light interference that has passed through a fiber-optic coil (about 5 km long). Two laser beams send one fiber but in the opposite direction. The beam moving against the rotation experiences a shorter path delay than the other beam. The resulting differential phase change is measured by interferometry, so a component of the angular velocity is converted to an interference pattern change measured in photometry.

Optical gyroscopes do the same thing as electromechanical gyroscopes, except that they do not have a moving part and are more accurate due to a phenomenon called the "sinak effect." Named after the French physicist Georges Sainak, Sinak's work is an optical phenomenon rooted in Einstein's theory of general relativity. In this phenomenon, a beam of light is refracted into two beams. Then the two beams travel in a closed circuit in two different directions and finally meet in a single detector. The light travels at a constant speed, so the device's rotation and the path that the light must travel will cause one of the two beams of light to reach the detector sooner. If there is a ring with the following structure in the direction of each axis, this phase change, known as the "Sinak effect," can be used to determine the direction.

The tiniest high-performance optical gyroscope ever built is no more significant than a golf ball and, therefore, unsuitable for many portable devices. The smaller the optical gyroscope, the weaker the signal, making it harder to detect motion. So far, this problem has not allowed optical gyroscopes to be made in smaller sizes. Engineers at the California Institute of Technology, led by Ali Haji Miri, a professor of electrical engineering and medical engineering at the School of Engineering and Applied Sciences, have built a new optical gyroscope 500 times smaller than its ultra-modern counterparts. This optical gyroscope can detect phase shifts that are 30 times smaller than the phase shifts of similar systems. Engineer Haji Miri's new optical gyroscope owes its excellent performance to a new technique called "bilateral sensitivity enhancement." Here, "bilateral" means that both rays of light in the gyroscope are affected. Sinak's effect distinguishes between two rays moving in opposite directions, which is why this phenomenon is considered non-reciprocal. Inside the gyroscope, light travels through tiny optical fibers. However, defects in this light path (such as temperature fluctuations or light diffraction) may affect the beams. Also, any other external interfering factor will affect both beams.

4.3.3 INERTIAL MEASUREMENT UNITS (IMUs)

Inertial measurement units (IMUs) and attitude and heading reference systems (AHRS) are produced in large quantities to measure orientation parameters. The primary target is to fuse accelerometers, gyroscopes, and magnetometers with temperature-calibrated data and an AI-based fusion algorithm to provide precise and trustworthy orientation. IMU benefits high performance, trusted reliability, and high quality. This sensor uses MEMS gyro and accelerometer technology to

produce precise, compact, and lightweight devices. It also provides direct mea
surement of acceleration and angular rate. It is entirely temperature-compensated
and calibrated over the operating temperature. Selecting the appropriate sensor for
various applications, such as construction and agricultural vehicles, unmanned
ground vehicles (UGVs), and robotics, is made possible by various performance
levels and packaging possibilities.

An inertial measurement unit, or IMU, is an electronic device that uses a
combination of accelerometers, gyroscopes, and sometimes magnetometers to
measure and report body-specific force, angle, and orientation. IMUs are commonly used to maneuver aircraft (orientation reference systems), including
drones (UAVs), UGVs, UUVs, and various other systems such as spacecraft,
satellites, and landers. IMUs are also successfully applied in robotics, unmanned
vehicles, and off-highway vehicles. Industrial robots are increasingly employed
in many sectors and applications because they can be taught to carry out hazardous, unpleasant, repetitive, or other jobs with consistency, accuracy, and precision. An industrial robot that uses Microstrain IMU can compensate for terrain in
real time and maintain stability. UGVs' ability to execute depends on their navi-
gational subsystems. Conventional, high-performance navigation systems support
unmanned solutions, but their use is constrained by cost, size, weight, and power
consumption issues. Unmanned ground vehicles may reliably navigate
unstructured environments with the help of microstrain IMUs, thanks to their low
power and economical design. Construction, mining, and agriculture vehicles are
subjected to unpredictable operating circumstances. For maintenance procedures
to be optimized and uptime to be maximized, it is crucial to understand the exposure of specific vehicles. Microstrain's IMU offers dynamic inclination detection, real-time vehicle stability management, and terrain adjustment in a
ruggedized package to give the vehicle the necessary information reliably.

Recent developments have made it possible to produce GPS devices equipped
with IMUs. The IMU allows the GPS receiver to operate when GPS signals are
unavailable, such as in tunnels, interiors, or under electronic interference. A
wireless IMU is known as a WIMU. Inside IMUs, an inertia unit detects linear
acceleration using one or more accelerometers and rotational speed using one or
more gyroscopes. Some also have a magnetometer, commonly used as a heading
reference. Typical settings include each accelerometer, gyroscope, and magnetometer on each axis for each of the three main axes: pitch, roll, and yaw.

An IMU is used to calculate the vehicle direction and to head associated with
magnetic north. The data collected from the IMU sensors allows the computer to
track the location of vehicles using a method called dead-end auditing. For
inland vehicles, an IMU can be integrated into GPS-based vehicle navigation
systems or vehicle tracking systems, giving the system the ability to calculate
and collect accurate data on the vehicle's current speed, rotation speed, direction,
inclination, and acceleration. Combined with the wheel speed sensor output, if
present, it gives a reverse gear signal for better traffic analysis.

In addition to navigation purposes, IMUs act as orientation sensors in many
consumer products. Almost all smartphones and tablets have an IMU as an

orientation sensor. Fitness trackers and other wearables may include an IMU to measure movement, such as running. The IMU can also determine individuals' growth levels while moving by identifying the specificity and sensitivity of specific running-related parameters. Some gaming systems, such as the remote control for the Nintendo Wii, use the IMU to measure motion. Low-cost IMUs have made it possible for the drone industry to thrive. They are also often used for sports technology (technique training) and animation programs.

4.4   SAFETY SENSORS

These scanners are the safety sensors that increase the robot's efficiency, speed, safety, and decision making. In intelligent agricultural implements, these sensors play a significant role as a protector of the device, people, and surrounding objects to prevent personal and objective damage. The development of safety sensors in recent decades has attracted much scientific attention. Various industries benefit from these sensors, including the food industry, biotechnology industry, medicine, environmental control, automotive industry, and robotics in all related industries due to their ability to obtain a rapid, selective response, identification, and analysis of the environment is very useful. The central part of a safety sensor is the element of that sensor. The sensor element is in contact with a detector. This element is responsible for identification and investigation in a mechatronic system. The detector then converts the signals generated from detection by the sensor element into a measurable output signal.

In AAVs, in addition to the positioning and orientation sensors, safety sensors are also necessary to control and react to obstacle detection in their safety zone. An AAV without a safety control can be a dangerous system. For this reason, the third group of basic sensors of agricultural robots that detect the presence of obstacles is named safety sensors. An autonomous robot must have a safety zone to stop moving if humans and any obstacles enter this area. Creating such a safety zone also prevents robots from colliding with other devices, trees, and buildings. This zone can be covered using safety sensors such as laser scanners (two-dimensional and three-dimensional), cameras (CCD, multifaceted, omni-directional, etc.), and switches (shock switches, bumper switches, emergency switches, etc.). These sensors scan the safety zone of robots and stop them if necessary. The following sections explain and classify the safety sensors based on their active principles, including laser-based, wave-based, vision-based, sound-based, and contact-based.

4.4.1   WAVE-BASED SAFETY SENSOR

Some safety sensors work on waves, such as infrared, laser, etc. In these sensors, the farther away from the light source, the more difficult it is to detect small objects. The act of utilizing lasers to collect point clouds in 3D spatial data is known as laser scanning. It can accurately measure and gather data from objects, surfaces, and landscapes. Laser scanners collect data as point clouds consisting of millions of three-dimensional coordinates (X, Y, Z coordinates). Laser scanning reduces data-collecting errors and the requirement to return to the site in case any information is missing. It mainly maps topography and produces 3D scans (or models) of building elevations, floor plans, tunnel profiles, and rail and highway bridges. Laser scanning is an efficient technique for accurate 3D data and model creation. Once cleaned up, the data from a laser scanner can be used in various 2D and 3D applications. Some companies, such as Leica, Trimble, Faro, Riegl, and Topcon, develop various application products. Laser scanners can be classified into three categories based on the technology utilized. While all scanners operate on the same basic premise of emitting laser pulses, they differ in how they seize the returned signals. Table 4.3 presents the different types of laser scanners.

The time-based scanner determines each point in 3D coordinates based on the distance measured for the region scanned. A laser scanner can record 1 million points with their corresponding in one single scan. The digital is created by processing the laser scanner's point cloud. Phase-based laser scanners operate on the same principles as time-of-flight scanners in that they emit a constant stream of laser beams and measure the phase shift to determine distance. This scanner is far more accurate than a time-based one and has a 300-meter scanning range. This scanner is used in various projects, including heritage, archaeology, architecture, and civil engineering. It is more flexible than a time-of-flight scanner. Triangulation laser scanners scan objects that need to be scanned with micron-level accuracy. These are often utilized for brief scans between 0.5 and 2 m. A triangulation laser scanner uses trigonometric calculations and has three

TABLE 4.3
Different Types of Laser Scanners and Performance Comparisons

Type	Mechanism	Specific	Operation Formula	Pros. and Cons.
Time-based	The same principle of a laser range finder	Measures the distance from an object based on time	Distance = (Speed of Light × Time of flight)/2	High scanning time and covers a large area
Phase-based	Computing a "shift" or "displacement"	Determines an object's distance by the difference in phase shift	Time-of-Flight (TOF) = Phase Shift/(2π × Modulation Frequency)	High reliability
Triangulation	To scan objects that require micron-level detail	The trigonometric calculation comprises three elements: a laser scanner, a camera, and the object	Trigonometric calculation	Low price and is famous for its precision level and inability to digitize reflective or transparent surfaces

primary components: a camera, a laser scanner, and the object to be scanned, which is mounted on a revolving plate to capture various faces.

The small 2D laser scanner is one of the ideal sensors for AAVs. These sensors can provide sufficient safety even at high speeds without compromising the device's efficiency. One of the advantages of these sensors is their low price compared to 3D scanners, which reduces the cost of farming robots. In contrast, 3D scanners are scanners that serve as scientific tools for detecting and analyzing the shape of objects in the real world or work environment of farming robots or for data collection (such as color, surface, properties, etc.) around the robot. They are often used to perform 3D reconstruction of the environment to create digital models of natural objects. These models have various applications, such as commercial design, defect detection in fabricated tools, reverse engineering, medical information, biological information, criminal identification, film production, 3D creative games, etc. Despite the widespread use of these sensors, high cost and complex and expensive object reconstruction algorithms are among the obstacles that make their use less common in everyday applications of agricultural robots. Currently, there is no one-piece reconstruction technology, and the methods are often designed and implemented according to the object's surface characteristics.

The following sensor in this range is an infrared sensor. This sensor performs the detection process by sending invisible infrared light rays. A receiving part detects any reflection of infrared light sent by the transmitter. These reflections allow infrared switches to determine if an object is nearby. Since these switches are only excited, they modulate them to more complex optical switches with a particular frequency and have receivers that respond only to that frequency.

4.4.2   VISION-BASED SAFETY SENSORS

The next group of safety sensors works based on images. These sensors are used when the environment gets complicated, or a complex decision must be taken. The detected data from these sensors are significant, and the analysis of this data takes longer. Vision sensors are excellent for reading codes and precise compartment location. The sensors detect the markers in single- or double-depth shelves for compartment fine positioning. The camera-based gadgets reliably recognize 1D or 2D codes on objects during code reading. The vision sensors include a video camera, display and interface, and computer processor. These are frequently employed in measurement, pass/fail judgments, and other observable aspects of product quality. An intelligent camera has a CPU built right into the device.

Vision sensors' cameras are exposed to the manufacturing process and are therefore vulnerable to corruption or damage. Hardened camera covers and lenses are available to guard against camera theft. Most vision sensors include permanently attached cameras to guarantee that the camera records the right field of view. Camera positioning is acceptable using brackets, arm mounts, and vibration-resistant mounts. Usually the camera and support only need two points of connection. Computer workstations are kept away from machinery in manufacturing environments to avoid risk to the costly interface. These include

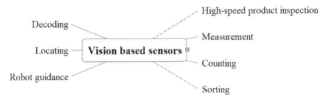

FIGURE 4.10   The advantages of the vision-based sensors.

lockable cabinets, translucent housings made of acrylic or polycarbonate, and dust-resistant designs. Wired connections, such as Camera Link, FireWire, USB, Ethernet, or composite cable, make communicating easier for the camera and computer. Figure 4.10 presents the advantages of vision-based sensors.

Data exchange between a video camera and computer-based software is made possible by machine vision systems. The operator compares the acquired photos to the criteria to decide the following action. Barcodes, blot/stain detection, size and alignment, and other properties identified entirely by non-contact examination can all be used as criteria for this selection. Vision sensors are advantageous when a product needs to have several features examined. For each piece, there is typically a tolerance range. Vision sensors are far faster and more precise than humans performing the same activity. While there are many advantages to machine vision sensors, there can also be severe drawbacks. Unforeseen events and input are too much for machine vision algorithms to handle. High development expenses for installation and employee training are anticipated, although ultimately saving money. Maintaining consistent illumination levels can be challenging, and cameras may have trouble isolating objects in crowded spaces.

4.4.3   Sound-Based Safety Sensors

The sound sensor is one module used to pick up on sound. This module is typically used to measure sound intensity. This module is mainly used for switch, security, and monitoring purposes. For the convenience of use, the precision of this sensor can be altered. This sensor uses a microphone to send input to a buffer, peak detector, and amplifier. After processing, this sensor detects sound and sends an o/p voltage signal to a microcontroller. It then does the necessary processing.

Like human ears, sound detection sensors have a diaphragm that transforms vibration into signals. However, a sound sensor differs because it has an amplifier (LM386, LM393, etc.) that is extremely sensitive to sound and has an integrated capacitive microphone, peak detector, and other components.

These parts make it possible for the sensor to function:

- Sound waves move through molecules of air.
- These sound waves cause the microphone's diaphragm to shake, which changes the capacitance.
- The processing of sound intensity then involves amplifying and digitizing the capacitance change.

TABLE 4.4

Different Types of Sound Sensors and Technical Comparison

Type	Mechanism	Specific	Pros and Cons
Carbon Microphone	The resistance of carbon granules	Electrical contact	A battery is needed
Ribbon Microphone	Corrugated aluminum ribbon	Moving by the velocity of the air	No active components, and famous for high-quality audio
Dynamic Microphone	Made up of a magnet and a diaphragm	Free-standing microphone	Cost effective
Condenser Microphone	Consists of a thin membrane nearby	Electrically conductive	Most expansive frequency response and a high quality

Other commonplace applications for sound sensors include:

- Consumer electronics like phones, computers, and music systems.
- Systems for security and monitoring, such as door alarms and burglar alarms.
- Home automation, such as sensing whistles or claps to turn on the lights instead of physically turning the switch.
- Recognition of background noise and sound intensity.

Many security incidents now start with sound, such as gunshots, violent behavior, or smashing glass. However, cameras with built-in sound exposure capabilities can significantly improve the security system because they automatically send out alerts when events, both actual and potential, happen. They then initiate prompt and appropriate efforts to lessen the effects. The sound sensor module is described in general detail on this page. A diaphragm is a component of many sound sensors that transforms air vibrations into electrical impulses. The sound sensors can detect noise levels within the range of frequencies humans can perceive. Different types of sound sensors are presented in Tables 4.4.

4.4.4 Contact-Based Safety Sensors

Proximity sensors or switches can detect the presence of adjacent objects without any physical contact. An adjacent sensor often emits an electromagnetic field or electromagnetic radiation (for example, infrared) and looks for changes in the field or return signal. The object the sensor senses is often called the proximity sensor's target. Different proximity sensors are needed for different purposes. For example, a capacitive proximity or photoelectric sensor may be suitable for plastic purposes. An induction proximity sensor always needs a metal target. Proximity switches are used in many different manufacturing processes. Some examples include measuring the position of device components, security systems, programs such as door opening detection, and automation that can monitor and guide the robot or its proximity to objects. Different proximity switches include infrared, acoustic or ultrasonic, capacitive, and inductive.

4.4.4.1 Ultrasonic Proximity Switches

These sensors are similar to infrared models but use sound instead of light. They use a transducer to transmit inaudible sound waves in a predetermined sequence at different frequencies. The time it takes for the sound to hit an object nearby and return to the second converter is then measured. In essence, sound proximity sensors measure the time it takes for sound pulses to "echo" and use this measurement to calculate distance, just like sonar used in submarines.

4.4.4.2 Capacitive Proximity Switches

These switches detect distances from objects by detecting changes in the surrounding electrical capacity. A radio frequency oscillator is connected to a metal plate. When the screen approaches an object, the radio frequency changes and the frequency tracker sends a signal to open or close the switch. These proximity switches are so sensitive that they are more sensitive to electrified objects than those not. Capacitive proximity switches detect metals and non-metallic materials such as plastics, glass, water, and oil. When non-metallic objects are discovered, the corresponding detection distance is affected by parameters such as conductivity, dielectric constant, and water absorption rate of the object. The detection distances are different, and the maximum detection distance for base metal conductors has it.

4.4.4.3 Induction Proximity Switches

These switches sense the distance to objects using magnetic fields. They are similar to metal detectors. The wire coil is charged with electric current, and an electronic circuit measures this current. If a metal object is close enough to the coil, the current increases, and the adjacent switch opens or closes similarly. The main disadvantage of these switches is that they can only detect metal objects.

4.4.4.4 Emergency Switches

The emergency stop buttons/switch are wired in series with the control equipment circuit of the machine equipment. The circuit breaks the vehicle and shuts the power supply by pressing the emergency switch. Some systems require a key that is not reversible. These switches are in the normal close or NC, which means that this switch is closed by default. But the difference is that this key does not return automatically, and when we put our hand on it, it does not return to its previous state by itself; it locks, and we have to turn it to return the key to its original state.

5 Communications

5.1 COMMUNICATIONS

The sending and receiving of data for an AAV is an essential act of communicating between internal components of the system and external vehicles, devices, and control units. Each AAV has communication systems between internal components (internal communication systems) and other communication systems by which they communicate with other robots (external communication systems). Internal communication systems are always used to connect two/more electronic parts of a robot at close distances, such as connecting a computer to an ECU. This communication can be one-way, two ways, or bus communication. Also, each AAV may need to send or receive data from an external source. External communication systems establish one-way or two-way communication between two AAVs or one AAV versus controllers, cloud(s), monitoring systems, user interfaces, or emergency switch. An example is a connection between the main tractor robot and a follower tractor robot. In the following sections, we will present more detailed and specific knowledge about the communication of AAVs (Figure 5.1).

5.2 INTERNAL COMMUNICATION

5.2.1 DIRECT CONTROL SYSTEMS

In this section, all communication systems between different components and parts within a farming robot are known as the internal communication system. Using these systems has gone through various stages over time. At the beginning of developing agricultural robots or implements, these instruments did not have a central control circuit or ECU. For this reason, there was no electronic communication between the agricultural device and the computer. It meant that the development of a robot and the control of various components of an agricultural vehicle, such as a tractor, required the installation of various electronic components. The device had to be equipped with a computer, and the rest of its controllable components were connected to actuators such as electric motors, electrostatic, hydraulic cylinders, etc. It was necessary to provide a communication system for the computer to communicate with the operators. At this time, all operators had to be controlled directly by a computer. This type of control system was called a "direct control system."

As you can see in Figure 5.2, this type of system was not a communication system, and more, because of the structure and function, control operations are used. IO ports were used for AD, DA, or PMC boards. IO ports were used to receive information from switches and control operators and relays (magnetic

DOI: 10.1201/9781003296898-5

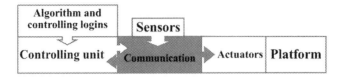

FIGURE 5.1 Communication in the autonomous agricultural vehicles.

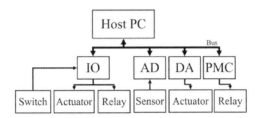

FIGURE 5.2 Configuration of direct control systems.

switches) using a command sent from a computer or a signal sent from switches and other sensors. AD boards were mounted on the computer bus to receive sensor output information. The DA and PMC boards-controlled actuators and motors, respectively.

5.2.2 SERIAL COMMUNICATION

After the development of ECUs and their installation on various devices, the communication between the computer and the sensors and the computer with the operators was no longer done directly through IO ports, AD, DA, or PMC ports but through control signals. The computer's output was sent to the ECU, and the operators were controlled via the ECU. Of course, more than external sensors such as positioning or orientation sensors could be connected to the computer, but the primary sensors of devices, such as engine speed sensors, etc. send data directly to the ECU, and the leading operators of devices such as brakes, etc. directly from ECUs were commanded. At this time, many serial communication ports were accepted by design engineers. The RS232c port was used for low-speed communications, and the RS422 and RS485 ports were for MSBs, cables, and later USBs to increase communication speeds. Figure 5.3 shows that the RS232c port can only connect two components and the system, while the RS422 and RS485 ports can only connect 10–32 devices. The RS232c port has limitations, such as restrictions on cable length, low data transfer speeds, and restrictions on multiple communications. These restrictions have prompted researchers to look for an alternative method since 1996 when the USB port was a viable alternative. The USB port could be an excellent alternative to the significant problems of the RS232c port. Still, as you can see in the Figure 5.3, this port had a master-slave protocol (master means the central or commanding system, and slave means accepting the commanding system). So in using this port, the slave could not do

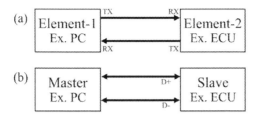

FIGURE 5.3   The architecture of (a) RS232c and (b) USB ports.

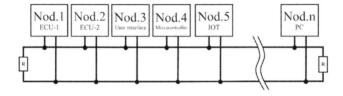

FIGURE 5.4   The architecture of CAN-bus.

anything without having a master or host system or even affect the master's commands. But the best advantage of this type of internal communication system is that the decision is made by the central computer and sent to the operators, increasing the data processing speed and reducing the data transfer speed.

5.2.3   CAN-Bus

In 1990, the ISO-bus standard was published, and since then, agricultural equipment communication systems have complied with ISO 11783 based on SAE J1939. This standard introduced the control area network, CAN-bus, or bus connection. Communication arrangements are required for robotic systems. As shown in Figure 5.4, using CAN-bus systems, various components and microcontrollers within a computer system can communicate without needing a computer or central system. In this case, the different ECUs of a system, computers, microcontrollers, and even the IoTs as nodes can communicate through bus communication. The need for a platform for multiple ECU connections may seem unnecessary. Still, in a robotic system, different ECUs or nodes should be controlled. The required controls in an AAV are: steering, forward, backward, brakes, shifting, torque, speed, power axis, and user interfaces. Even the central circuits of modern back and front devices require cross-communication and multiple communication platforms. The development process of an AAV's internal communication starts from direct control systems, upgrades to serial ports, and finally uses the CAN-bus to match.

5.3   EXTERNAL COMMUNICATION

Just as the devices and internal components of a farming robot interact, robots must be able to communicate with the control unit, monitoring unit, cloud

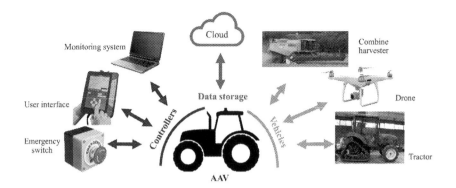

FIGURE 5.5   Illustration of external communication variation in AAVs.

servers, control tools, or other farming robots such as tractors, combine harvesters, drones, etc. (Figure 5.5). This type of communication requires wireless communication, which we have introduced in this book as external communication systems. This type of communication system can be two-way communication (either device, device, or system can receive or send commands), one-way (communication is sent only from one device, device, or system to another, and there is no feedback), or master-slave (one device or system issues commands and the other system operates according to commands). For example, two-way communication can be an interaction with two tractor robots using Wi-Fi signals and interacting with each other to send messages about the vehicle's position, the safe range of vehicles, and even the presence of obstacles in the path. For a one-way connection, the Bluetooth emergency keys in the operator's hands to stop the device in an emergency can be a good example. For master-slave communication, multi-robot systems can be used, for example, to operate in one or more agricultural lands using several tractor robots, one robot as a supervisor or master robot, and the other robots as robots. The slave function will work. The other robots will stop if the master robot stops until the next move. As shown in Figure 5.5, all communications between farming robots can be done via Wi-Fi, Bluetooth, or LTE signals and using Wi-Fi modems, Bluetooth gadgets, personal digital assistants, or mobile phones.

5.3.1   Wi-Fi

Wi-Fi is easy to use but it has its complexities and hassles. The Wi-Fi interface uses radio waves to transmit data according to IEEE 802.11 standards on the network. Wi-Fi routers or modems can exist as separate or integrated devices in ADSL, WiMAX, or other devices. An internal controller is responsible for managing signals, encrypting and decoding, and transmitting information; like any other radio device, the antenna will be an integral part of it. Most routers use the 4.2 GHz band for data transmission, which is used in Bluetooth interfaces and microwaves. That is because when the number of Wi-Fi devices increases in the 4.2-GHz band, the amount of interference in this band is high. By defining the required standards and regulations, it is possible to use the 5-GHz band, and the new routers can support two bands of 4.2 and 5 GHz simultaneously. Using a 5-GHz band minimizes interference, but Wi-Fi receivers should also be able to support it.

Prototypes of the Wi-Fi standard used only the WEP encryption system, which was easily breakable and permeable. Over time, WPA and WPA2 encryption systems were introduced, which increased security. The use of cryptographic systems takes up part of the bandwidth, but it is worth increasing the security and protection of the information. In robotic systems, control security is one of the topics discussed in international forums. The WPS feature, introduced in 2007, eliminates the need for the user to enter a password to connect to the network and press a button to exchange passwords and connect to the wireless network. However, it is said that this feature has low security, and a hacker can penetrate the network through it in a few hours.

Wi-Fi antennas come in various types and are designed for indoor or outdoor use. To design a farming robot, you must remember that the antenna used must have a high range so that the exchange of information between the robots can be done efficiently. Directional antennas transmit the signal in only one direction and to another device in the signal transmission path. Omnidirectional antennas are found in most home and office routers and transmit the signal from one point to several directions. Wi-Fi antenna coverage depends on various factors, including the output power of the router or wireless network card, the ability to receive signals, various environmental barriers to data transmission, and so on. Output power is measured in milliwatts, which can vary depending on the type of network card used (the higher the value, the higher the output signal strength). Antennas centralize the received signal and can amplify the output power. The most important number that is always considered about Wi-Fi antennas is their decibel rate, which means that the higher it is, the more likely it is to send a signal to a farther space.

Using the MIMO method is one of the solutions used to increase the antenna and control the robots. This method uses several antennas, inputs, and outputs in the radio transmitter and receiver to increase data transmission efficiency. Of course, the words "antenna" and "multiple inputs and outputs" in the MIMO definition do not necessarily mean that there are multiple physical antennas in the router but instead refer to data transmission channels. This technology can increase the data transfer speed without needing more power or bandwidth, which the manufacturers of Wi-Fi devices have considered. It works by creating data transmission power on different channels and antennas, creating votes of information for transmission, and improving stability and efficiency. These have led to the inclusion of the MIMO standard in 802.11n, 4G, WiMAX, and HSPA + connectivity.

5.3.2 Bluetooth

Bluetooth is a short-range wireless technology that allows mobile phones, PDAs, stereo recorders and players, home appliances, cars, and all other devices to connect. The original idea for Bluetooth came from Ericsson Mobile in 1994. Ericsson, a Swedish telecommunications company, was building a low-cost, low-cost radio connection between a cell phone and a cordless phone. Engineering began in 1995, and the original idea went beyond cell phones and handsets to include all mobile devices. It is a standard for radio waves used for wireless communication on portable computers such as laptops, cell phones, and standard electronic devices. These waves are used for short distances and are considered low-cost technology for wireless communications. And with this technology, you can exchange messages, photos, or any information wirelessly between two devices that have this technology. The Bluetooth radio is housed inside a microchip and operates in the 2.4-GHz frequency band. This technology uses the "Frequency Hoping Spread Spectrum" system, whose signal changes 1,600 times per second, which is a great help to prevent unwanted and unauthorized interference. In addition, the identification code of the other party is checked by the software. It will ensure that your information reaches only the intended destination. These waves exist with two powers. A lower power level can cover small environments (for example, inside a room), or a higher power level can cover moderate suffering (for example, it can cover an entire house). This system can be used for both point-to-point and point-to-point communication. This technology uses the receiving and sending system in the right direction. It can pass radio waves through walls and other non-metallic obstacles. If the disturbing waves of a third party cause interference, the transmission of information is slow but does not stop. With today's systems, more than seven devices can be activated to communicate with the wave generator in another device. It is called a Piconet. Multiple piconets can be connected to form a scatter net.

Bluetooth is a wireless feature that defines short-range communications between devices equipped with small, dedicated Bluetooth chips. Bluetooth eliminates cables and provides a wireless way to connect computers to all electronic devices, creating small, private computer networks known as PANs or personal networks. Bluetooth creates a common language between different devices, allowing them to communicate and connect easily.

However, the short range and limited speed of Bluetooth make it less common for wireless LANs, as these computer networks are typically more than 10 m away from Bluetooth and have a speed range of 10–100 MB per second. Some devices you still use, such as the parking door controller or the latest generation of cordless phones, use ISM band frequencies. Ensuring that Bluetooth waves do not interfere with the waves of these devices is one of the most challenging steps in designing this technology. The essential advantages of Bluetooth devices is that they are wireless, low cost, and automatic.

5.3.3 INFRARED

There are other wireless communication methods, such as infrared or infrared communication. Infrared is light waves whose frequency is lower than the frequency that can be seen and understood by the human eye. Most remote-control devices with audio and video equipment use infrared to send information. Infrared communication and information transfer are reliable and secure and do not cost much. But there are two limitations to its use: first, that infrared radiation is only emitted in the direct path, and second, that infrared technology is a one-to-one technology. At the same time, it can only communicate between two devices. For example, you can transfer that information from your laptop to your mobile phone, but you cannot transfer that information to your friend's PDA at the same time. Of course, these two infrared features are considered advantages in some cases because the information transfer operation is located only between the two devices, and there is no possibility of interference with other devices, which is impossible. This feature allows your information to be sent only to the device you want, even if you are in a place full of infrared receivers. Bluetooth technology was invented to cover the limitations of infrared. The maximum data transfer rate on Bluetooth devices running the older Bluetooth 1.0 standard is 1 MB/s, but on the Bluetooth 2.0 standard, data can be transferred at 3 MB/s. Bluetooth devices that use the new standard are also compatible with older ones.

5.3.4 LTE

LTE stands for long-term evolution and is a registered trademark of ETSI (European Communications Standards Institute) for wireless data communication technology and the development of GSM/UMTS standards. However, other countries and companies actively participate in the LTE project. The goal of LTE was to increase the capacity and speed of wireless data networks using DSP (digital signal processing) signal processing techniques and modulations developed during the millennium. The next goal was to redesign and simplify the network architecture to an IP-based system with significantly reduced transmission time compared to the 3G architecture. The LTE wireless interface is incompatible with 2G and 3G networks, so it must be operated on a separate radio spectrum.

The LTE specification has a maximum rate of 300–75 Mbps and quality of service regulations that allow transmission latency of less than 5 MB on the radio access network. LTE can quickly manage mobile phones and support multimedia streams and broadcasts. LTE supports scalable carrier bandwidth from 1.4 to 20 MHz and supports both split frequency duplex (FDD) and temporal duplex (TDD). The IP-based network architecture, in which EPC is designed to replace the leading GPRS network, supports seamless handover of voice and data to cell towers with older network technologies such as GSM, UMTS, and CDMA2000. The simpler architecture reduces operating costs (for example, each E-UTRA cell supports up to four times the information and audio capacity supported by HSPA).

5.4 PORTS

In communication systems, various ports with different configurations can be used. A computer port is an interface or connection between a computer and peripherals. The primary function of a computer port is to act as a connection point. The communication ports are the connection point between two devices, such as a computer with a keyboard, mouse, external sensors, and displays. For example, some ports can connect sensors and actuators such as RTK-GPS, IMU, and robotic arm controllers. The classification of communication can be divided based on data type, port type, generation, and so on. In the following section, we ignored the classification and introduced commonly used posts in the past.

5.4.1 PARALLEL PORTS

Parallel ports connect computers with devices that need parallel data transmission. Data is transmitted quickly in or out of parallel using more than one line or connected wire. Parallel ports are widely used to connect parallel devices such as old-type personal printers and currently used industrial printers. Generally, it transmits 8 bits of data per time at a speed of 50–100 kbps (K bit/second). This feature is used for fast data transfer. These ports commonly have DB-25 (25-pin) connectors and 36 pins in centroids parallel 36-pin type. It was expected in old AAV and prototype versions and is now being replaced by USB (Figure 5.6).

5.4.2 RS PORTS

Serial ports are interfaces through which peripherals can be connected to a serial protocol by transferring 1 bit per unit of time to a communication line. The most common type of serial port is a D-sub, which transmits RS-232 signals. These types of ports are used to connect keyboards, mice, PLCs, UPSs, etc. Most serial ports use 9- to 25-pin RS-232C connectors, DE-9, and DB-25. In the 15th generation of computers, serial ports are less commonly used and have gradually been replaced by USB ports. The other RS posts are RS422 and RS485 (Figure 5.7).

5.4.3 PS/2 PORTS

IBM made this connector to connect the mouse and keyboard. Its purple and green colors were defined for the keyboard and mouse, respectively. The PS/2 is

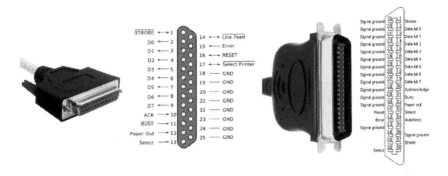

FIGURE 5.6 Types and structure of parallel ports.

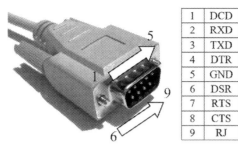

1	DCD
2	RXD
3	TXD
4	DTR
5	GND
6	DSR
7	RTS
8	CTS
9	RJ

FIGURE 5.7 Serial port.

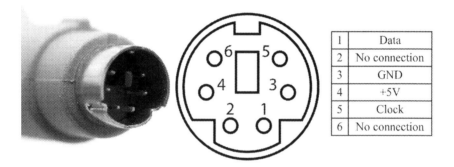

1	Data
2	No connection
3	GND
4	+5V
5	Clock
6	No connection

FIGURE 5.8 PS/2 port.

a 6-pin DIN connector. Of course, these types of connectors in newer systems have also given way to USBs (Figure 5.8).

5.4.4 AUDIO PORTS

Audio ports connect speakers or other audio outputs and devices, such as voice alerts of safety sensors. Audio signals can be analog or digital, depending on the type of port and its connection. One of the most common and practical connectors is the 3.5 mm TRS connector, which connects stereo headphones. This connector includes six connectors in blue, lemon, pink, orange, black, and gray. The other port in this group is the digital audio RCA plug style (Figure 5.9).

5.4.5 DVI PORTS

The DVI port is a high-speed digital interface between a controller such as a computer and a monitor. This port was developed to replace VGA technology and to transmit compressed digital video signals. Based on the type of transmission signals, DVI ports are divided into three types: DVI-I, DVI-D, and DVI-A. DVI-I

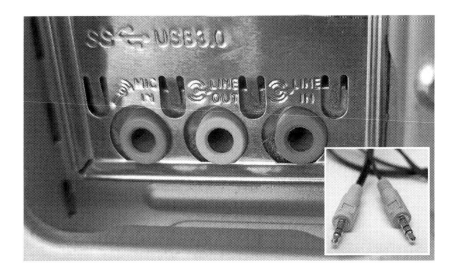

FIGURE 5.9 Types and structure of audio ports.

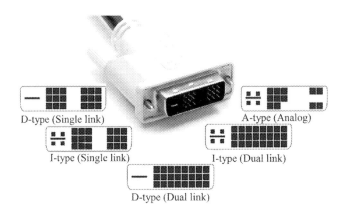

FIGURE 5.10 Types and structure of DVI ports.

is a DVI port for integrated analog and digital signals. DVI-D and DVI-A ports are designed to transmit digital and analog signals. Digital signals can have one or two link interfaces in which one link supports only one digital signal up to a resolution of 1920 × 1080, and a dual-link supports a digital signal up to 2560 × 1600. This group also has mini DVI, micro DVI, and LFH60 (dual DVI-D) (Figure 5.10).

5.4.6 HDMI PORTS

The HDMI port was a digital interface connecting computers to high-definition and high-quality devices such as PC monitors, high-definition cameras, and safety sensors. The HDMI port can be used to transmit uncompressed video

Standard
HDMI

Mico
HDMI

Mini
HDMI

FIGURE 5.11   Types and structure of HDMI ports.

signals and uncompressed and compressed audio signals. This port includes a
19-pin connector, and the latest version of this port, HDMI 2.0, can transmit
video signals with a resolution of 4096 × 2160 (Figure 5.11).

5.4.7   VGA PORTS

This port is used in most computers, projectors, video cards, and TVs. It is a
D-sub port with three rows of 15 pins called DE-15. The VGA port is one of the
primary interfaces between computers and older CRT monitors. Even newer
LCD and LED monitors have these ports, but using these ports reduces image
quality. This cable can transmit video signals up to 648 × 480 resolution. With
the development of digital devices, VGA ports have given way to HDMIs. This
port was present in older laptops which have been gradually removed in newer
systems. This type also has a mini-VGA (Figure 5.12).

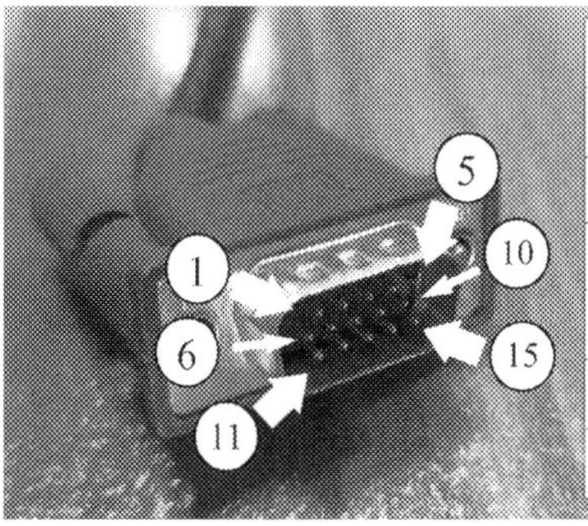

1	RED
2	GREEN
3	BLUE
5 ~ 10	GND
13	H-SYNC
14	V-SYNC
4, 11, 12, 15	Not connected

FIGURE 5.12   VGA port.

5.4.8   USB Ports

The USB port is an interface for replacing serial, parallel, PS/2 ports, or even portable device chargers. The USB port can be used to transfer data, as an interface for peripherals, and as a power supply for connected devices. There are three USB ports: Type A, Type B or Mini USB, and Micro USB. Type A is a four-pin port. Type-A versions have USB 1.1, USB 2.0, and USB 3.0 ports. USB 3.0 is the standard and supports data transfer speeds of up to 400 Mbps. USB 3.1 port supports data transfer speeds of up to 10 Gbps. USB 2.0 is black, and USB 3.0 is blue. The USB-C port is the latest version of USB ports and is a reversible, two-way connector. USB-C type is supposed to replace types A and B. This port contains 24 pins and can support up to 3 amps. This feature effectively transmits high current to the latest technology of fast battery charging in smart batteries and reduces battery charging time (Figure 5.13).

5.4.9   Ethernet Ports

An Ethernet port is a networking technology that connects a computer to the Internet, allowing various devices and robots to connect to different networks, computers, and devices. The RJ-45 port is one of the Ethernet ports. The user interface used for computer networks and telecommunications is known as the Registered Jack (RJ), and the RJ-45 port is used explicitly for Ethernet cable. The RJ-45 connector is an 8-pin, 8-pin (8P-8C) modular connector. The latest Ethernet technology is Gigabit Ethernet, which supports more than 10 Gbps data transfer speeds. Keyless 8P-8C modular connection is generally referred to as RJ-45 Ethernet. Most RJ-45 ports have two LEDs to indicate packet transmission and identification. RJ-11 is another Ethernet port used as an interface for telephone, modem, or ADSL. Although computers have seldom been equipped with an RJ-11 port, this port is one of the primary interfaces in all telecommunication networks. Although the RJ-45 and RJ-11 ports look similar, the RJ-11 is a minor port and uses a 6P-2C 6-point connector, even if a 6P-2C is sufficient (Figure 5.14).

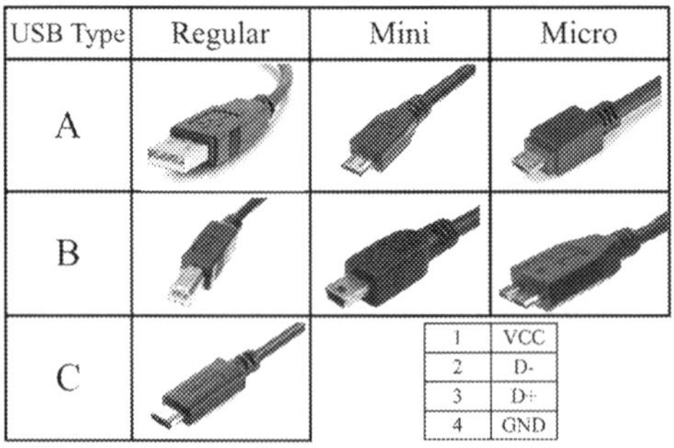

FIGURE 5.13   Types and structure of USB ports.

FIGURE 5.14 Ethernet ports.

5.4.10 S/PDIF/TOSLINK PORTS

This connection is an audio connection in the home media subset. This connection is the most widely used digital audio port that supports digital audio. Applications of this connection include home entertainment systems.

5.4.11 SATA PORTS

The e-SATA port is an AT external serial connection used as an interface to connect memory devices such as external hard drives. There is also a hybrid port

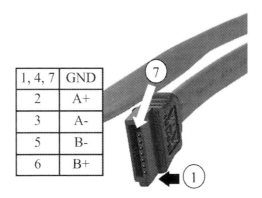

1, 4, 7	GND
2	A+
3	A-
5	B-
6	B+

FIGURE 5.15 SATA port.

that can support e-SATA and USB ports. None of the SATA and USB organizations have officially approved the E-SATA port, and using such ports can be dangerous and harmful (Figure 5.15).

5.4.12 Apple Ports

Apple also developed many specified ports that are less used in AAV. For example, Apple display connector – ADC, Apple high-density video HDI-45, Mac Video/MIDI/game port/AUI/DA-15, Apple desktop Bus-ADB, and Mac serial.

6 Controlling Units

6.1 CONTROLLING UNITS

Circuitry that controls activity in a computer's processing units is known as a control unit. When a control unit receives input data, it transforms it into control signals based on the related algorithm and sends them to the central processor. It informs the computer's logic unit, memory, and input and output devices on how to react to a program's commands. The computer's processor then instructs the associated hardware on what operations to perform. Figure 6.1 presents a schematic performance diagram of a control unit.

In an AAV, the controlling unit is a central segment that receives and sends related commands and signals to all components. It can be visualized as the robot's central brain and decision-making part. It gets all information from sensors via the communication interface, adapted with relevant algorithms, makes the necessary calculations, and sends data to actuators. Figure 6.2 presents the controlling unit configuration in agro-industrial.

There are many chances to choose AAVs' controlling unit, including controlling boards and computers. This chapter will discuss each in detail and present a performance comparison.

6.2 CONTROLLING BOARDS

Control boards are one of the first choices of CU in robot design. They are small, cost-effective, accessible to code, and mostly open source. Also, these boards are very suitable for muli-ECU and CAN-bus-based systems. A control board may be established early on in the project to monitor and oversee the operations related to parts, materials, and processes and create the program's parts, materials, and processes selection list. This list is typically created using a company's product line's approved parts list and approved materials and procedures list, which has been updated as needed and may be reorganized to meet customer delivery criteria [113]. This section has comprehensively discussed four controlling unit types: Arduino boards, Raspberry Pi boards, BeagleBone Black boards, and AdaFruit boards.

6.2.1 ARDUINO

The Arduino boards are with Linux technology and are open-source electronics platforms. The Arduino is built on simple hardware and software. A motor can be started, an LED can be turned on, and something may be published online

DOI: 10.1201/9781003296898-6

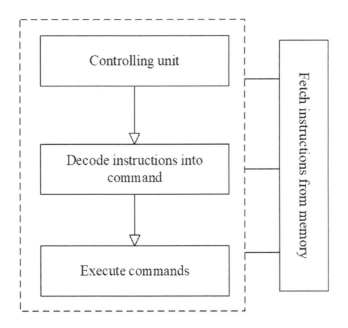

FIGURE 6.1   A schematic diagram of control unit performance.

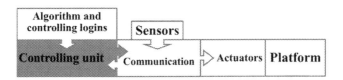

FIGURE 6.2   Controlling unit in the autonomous agricultural vehicles.

using an Arduino board to receive inputs like light on a sensor, a finger on a button, or a tweet. Boards, Lilypads, and shields are some of Arduino's products. Due to its flexibility, manufacturers and research enthusiasts can now create boards and shields specifically tailored to their research needs and application areas. Researchers and robot developers can use the Arduino open-source community to develop cutting-edge research applications and solutions ready for market in various fields, including home automation, robotics, wireless connectivity, drones, and many others [114]. Over the years, countless projects, ranging from simple household items to complex scientific instruments, have used Arduino as their brain. It attracted a large global community of makers, including students, hobbyists, artists, programmers, and professionals. Thanks to their contributions, this community has created a wealth of easily accessible knowledge that benefits both beginners and seasoned users. Figure 6.3 presents the main structure of the Arduino.

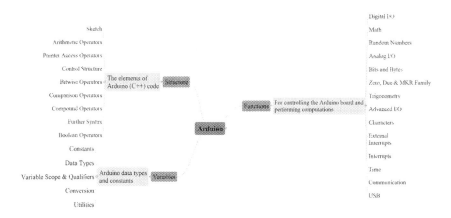

FIGURE 6.3   The structure of Arduino boards.

6.2.2   RASPBERRY PI

The Raspberry Pi is a fully functional credit card–sized computer that can run programs similar to a traditional desktop PC [115]. The Raspberry Pi is a small, inexpensive computer that connects to a computer monitor or TV and operates with a regular keyboard and mouse. With the help of this competent small gadget, individuals of all ages may learn about computing and how to program in languages like Scratch and Python. It has all the features of a desktop computer, including the ability to play high-definition videos, browse the Internet, create spreadsheets, word documents, and play games. Additionally, the Raspberry Pi can communicate with the outside world and has been utilized in various digital maker projects, including music players, parent detectors, weather stations, and tweeting birdhouses with infrared cameras. The company's aim is worldwide use to learn how to program and comprehend computers.

There are two models of the Raspberry Pi, including model A and model B. The USB port is the distinction between models A and B. Less power will be used by the model A board, which does not have an ethernet port. However, the model B board was created in China and had an Ethernet connector. Open-source technologies, such as communication and multimedia web technologies, are included with the Raspberry Pi. The Raspberry Pi board foundation introduced the computer module in 2014 to promote the usage of the model B Raspberry Pi board as an embedded system component. Figure 6.4 presents the Raspberry Pi hardware architecture.

The Raspberry Pi board includes RAM, a CPU, a GPU, an Ethernet port, GPIO pins, an Xbee socket, a UART, a power connector, and external device interfaces. In the case of mass storage, an SD flash memory card can be used. In the same way the computer boots into Windows from its hard drive, the Raspberry Pi board will boot from this SD card. The Raspberry Pi board's primary hardware requirements are an SD card with the Linux operating system, a U.S. keyboard, a monitor, a power source, and a video cable. Optional hardware requirements include a USB

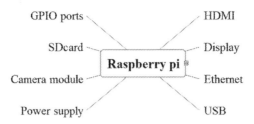

GPIO ports   HDMI

SDcard   Display

Raspberry pi

Camera module   Ethernet

Power supply   USB

FIGURE 6.4 The hardware architecture of the Raspberry Pi [116].

mouse, powered USB hub, case, and an Internet connection for either model A or B. For model B, a LAN cable is utilized for the Internet connection. The SD card in the board's slot serves as the Raspberry Pi's hard drive. A USB connector power it, and an HDMI port on a current monitor or TV can connect the visual output. It provides all of a typical computer's fundamental features.

6.2.3 BeagleBone Black

Art 15 of the FCC Rules is annotated on the BeagleBone to ensure compliance. The following two requirements must be met for operation: (1) this device cannot produce harmful interference, and (2) this device must tolerate all interference, including interference that might result in undesirable operation. The user's right to use the equipment could be revoked if changes or modifications are made without the express approval of the party in charge of compliance. This digital device falls under Class A or B and complies with Canadian ICES-003. The user's right to use the equipment could be revoked if changes or modifications are made without the express approval of the party in charge of compliance. This digital device complies with Canadian standard NMB-003. It is of class A or B. Changes or alterations that the party in charge of conformity did not expressly approve may have taken away the user's power to use the equipment.

The Beaglebone Black (BBB) is similar to a computer in that it comes in a small container with all the necessary ICs soldered to make a single circuit board, a processor, a graphic accelerator, memory, and other components. As a result, it is also known as a single-board computer. It makes use of the potent 1 GHz AM335x ARM Cortex-A8 processor. The Ethernet network, monitor, mouse, and keyboard may all be connected to this Beaglebone Black microcontroller board. This processor is booted using the Linux operating system.

Researchers mostly use this tool to create intricate projects and efficiently learn about the Linux Operating System. With additional functionality, this Beaglebone Black is identical to the Beaglebone. Compared to Beaglebone, it operates more quickly and is more commonly utilized. It is utilized in development-level automation, IoT projects, and robots [117]. The BeagleBone Black is incredibly easy to set up. It makes the BeagleBone Black the preferred option for projects that deal with electronics more directly and in a more complex way, along with the fact that it has 65 input and output pins and an absurdly large number of compatible ports.

On the BeagleBone Black, tasks like reading data from external sensors, controlling actuators (such as motors or lighting systems), and networking are more accessible and practical [118].

6.2.4 ADAFRUIT

An MIT engineer, Limor Ladyada, created Adafruit in 2005. She established an online environment for learning electronics and producing the best goods. Adafruit now employs over 50 people in its 15,000+ square foot manufacturing center in New York City. Adafruit has increased its selection to include hardware, software, and electronics [119]. Adafruit creates and distributes its development boards for educational and hobbyist uses and supplies third-party parts and boards like the Raspberry Pi. The business introduced a board containing an Atmel ATmega32u4 microcontroller and several sensors as the Circuit Playground in 2016; the more potent Circuit Playground Express followed it in 2017 [120]. The Circuit Playground Express was the best-selling item for Adafruit Industries in 2017.

Adafruit creates and manufactures Adafruit boards, which are microcontroller boards similar to Arduino. The boards were made with the Arduino platform in mind. Metro, Metro Mini, Trinket, Pro Trinket, Gemma, and Menta are Adafruit boards. Adafruit boards are, in theory, quite similar to Arduino boards and are likewise compatible with Arduino shields, although the company added several features that it considers crucial. Adafruit boards can be programmed using the Arduino IDE, just as Arduino boards. Table 6.1 presents the performance comparison of different types of boards.

TABLE 6.1
Performance Comparison Of Different Types of Control Boards

Type of Board	Advantages	Disadvantage
Arduino	Inexpensive, cross-platform, simple, clean programming environment, open source, and extensible software	Dumbing of the AVR microcontroller, cannot run more than one program at the same time, lack of built-in communications, and limited number of IDEs
Raspberry Pi	Small, working as a standard computer, relatively low price, low-cost server, and more cost effective than a regular server	Windows and Linux distros are not compatible, some applications on CPU processing are off-limits,
BeagleBone Black	Interaction of a multitude of external sensors and the creation of creating an Internet of Things	Lack of USB ports and lack of video encoding
Adafruit	–	–

6.3   COMPUTERS

A computer is a programmable electrical device. After executing logical and mathematical processes, it renders output and can save it for further use. Both numerical and non-numerical calculations can be processed by it. The Latin word "computer," which means to calculate, is where the word "computer" originates. A computer is made to run applications and offer a range of solutions through integrated hardware and software components. It functions with the aid of programs and uses a string of binary digits to represent decimal numbers.

Additionally, it features memory for storing information, software, and processed data. Hardware refers to the physical parts of a computer, such as wires, transistors, circuits, and hard drives. On the other hand, software refers to data and programs [121]. Figure 6.5 presents the different types of computer categories [122].

6.3.1   Microcomputers (Personal Computers)

A microcomputer is a fully functional computer that is tiny enough to be used simultaneously by one person. A microcomputer is commonly referred to as a single-chip microprocessor-based device or a personal computer (PC). Desktop and laptop computers are typical examples of microcomputers. Microcomputers are devices other than traditional computers, such as calculators, smartphones, laptops, workstations, and embedded systems.

Smaller than a mainframe or minicomputer, a microcomputer has a central processing unit incorporated into a semiconductor chip (CPU). They also come with input/output (I/O) ports, a bus or system of connecting cables, memory in the form of read-only memory (ROM), and random-access memory (RAM), all in one device. They are formally known as the motherboard. Keyboards, displays, printing devices, and external storage are standard input and output devices. Figure 6.6 presents the different types of microcomputers.

Nowadays, developers prefer building software and applications for cell phones rather than the microcomputers typically used to create them. In essence, users can fix simple computer issues and assemble and replace all hardware components of

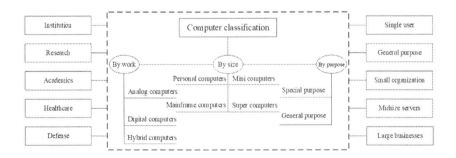

FIGURE 6.5   Classification of computers.

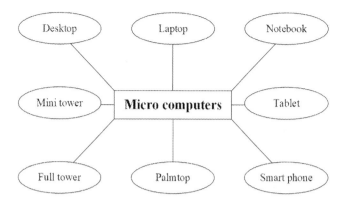

FIGURE 6.6 The different types of microcomputers.

microcomputers. Working with computers in an office environment does not require highly skilled personnel. Microcomputers are particularly helpful for marketers and students since they can be used to research, retrieve important information from the internet, and store that knowledge on other software loaded on the computer.

Each user is connected to the other by voice, video, text, and email using microcomputer equipment. You can post your remarks online on several websites, such as Facebook. Due to their compact size, specific microcomputer devices are portable and may readily transfer from one location to another. Most people can afford to acquire computers because of their low prices. A microcomputer enables a multi-tasking environment, making it possible for one device to perform several tasks at once, such as printing and scanning any document, using the Internet to browse, purchase tickets, watch movies, create games, and more. Microcomputers enable consumers to save money by completing complicated calculations faster and more accurately. Microcomputers are modest in size compared to larger computers like mainframes and supercomputers.

6.3.2 MINICOMPUTER

Because of the manufacturer's careful coordination of the parts utilized, a mini-PC (Mini PC) or small computer combines strength, proper performance, and part coordination to offer the user extremely high efficiency. Mini PCs are smaller and consume less power than personal computers and L-Ones, but they also give the user more workspace and require less maintenance due to the lower cost of electricity. It should be noted that placing the hard drive and RAM and installing and configuring the product both take less than five minutes. Additionally, features like the support for multiple displays, the existence of many ports, and the use of high-speed memory types like SSD and M2 memory point to a strong and effective product in the world of technology. These powerful portable computers give you processing power and speed wherever you go. The following uses for mini-PCs are

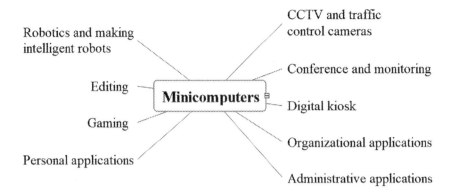

FIGURE 6.7   The different applications of minicomputers.

possible based on each model's technical requirements. Figure 6.7 presents the different applications of the minicomputers.

One benefit of small computers is that most lack ventilation and cooling systems. Minicomputers are incredibly effective in industrial settings thanks to this characteristic. Minicomputers can now be linked to the monitor's back thanks to their reduction in size. Due to the usage of low-power processors, minicomputers now typically run on 65-watt adapters due to their much-reduced power consumption. Nowadays, the majority of governmental and non-governmental organizations use small computers extensively. IT administrators now prefer minicomputers due to their long life span, low power consumption, lack of a fan, high speed, and small size.

6.3.3 MAINFRAME COMPUTERS

Computer systems known as mainframes often have a lot of storage, powerful processing, and high levels of dependability. Large enterprises primarily employ these computers for crucial applications that demand intensive data processing. Mainframes typically have unique skills (Figure 6.8). Mainframes still serve as the foundation of contemporary business in various industries, including banking, finance, public and government services, private firms, etc. Large businesses employ mainframes for critical applications requiring high processing speeds; the

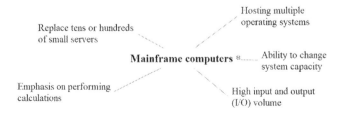

FIGURE 6.8   Capabilities of mainframe computers.

primary distinction between mainframes and regular computers is their software environment. As a result, mainframes are far more suited for numerical calculations for financial operations, comparisons, etc., because these systems' resources are more capable and were designed for this usage. It is preferable to define mainframe systems as systems for hosting applications that need more powerful resources than standard computers if we wish to be more specific. In essence, mainframes have hardware far more competent than typical personal computer systems, and this more potent hardware was explicitly designed for them to run programs requiring intensive processing. A supercomputer concentrates all of its power on running a small number of specific programs to process them as rapidly as possible may be the significant distinction between a supercomputer and a mainframe. However, the mainframe divides up its processing capacity among the concurrent execution of numerous programs.

6.3.4 SUPERCOMPUTER

Currently, the most powerful computers are supercomputers. Supercomputers are utilized in many fields, such as quantum physics, weather forecasting, oil and gas exploration, molecular modeling, physical aerodynamic simulations, nuclear research, and cryptanalysis. They are primarily optimized for executing specific scientific and technical computations, in contrast to mainframes, which may be employed for various activities.

With businesses of all sizes embracing this new technology, cloud computing is expanding more rapidly. Industry insiders predict that this trend will only intensify and spread over the following few years. Even if cloud computing is unquestionably advantageous for medium- to large-sized firms, it also has advantages, particularly for smaller businesses. Each supercomputer's early operating systems were created to boost speed. There are just a few examples

TABLE 6.2

Performance Comparison of Computers

Type of Computer	Advantages	Disadvantage
Microcomputer	General purpose, portable, multi-task, and proper size	Slow computing power, low memory size
Minicomputer	Low cost, small size, flexibility, portable, low energy consumption, and low noise	Lack of optical drive, small screen size, low memory size, lack of upgradeability, and unique operation system
Mainframe computer	High processing capability	Cost, setting up, size, maintenance, and environmental limitations
Supercomputer	Economic for huge tasks, almost unlimited storage, low processing time, and easy backup and restore	Technical issues, security in the cloud, and prone to attack

of the computationally intense scientific and engineering applications that su-percomputers are utilized.

For you to properly comprehend the notion of cloud computing with the aid of these institutions, we are now providing you with a list of cloud computing's benefits and drawbacks. Supercomputers are designed to carry out exceedingly complex calculations quickly. They do trillions of calculations every second and have tens of thousands of processors. Because supercomputers' processors ex-change information quickly, they provide lag-free work and are superior to distributed computing systems. Chess games, graphics-intensive displays, and precise weather simulations are all possible with these supercomputers.

Supercomputers process information in parallel, whereas conventional com-puters serially or sequentially do so. In other words, supercomputers perform multiple tasks concurrently rather than sequentially, as do regular computers, one at a time. Table 6.2 presents the performance comparison of different types of computers.

7 Performance

7.1 INTRODUCTION

There are many initiatives meant to solve the many environmental issues, and there are also several ways to evaluate them due to the wide variety of environmental issues. Environmental performance indicators are the techniques for evaluating the environmental effects and the usage of natural resources (both living and non-living). A project's development and success are typically gauged using performance indicators in various disciplines, including marketing, economics, education, and legal studies. A person's actions can be evaluated by specific indicators, while others can capture the efforts of entire nations or even the entire world. In particular, performance indicators of robotic systems in agriculture try to evaluate the quality of operation in a field [123,124].

The success or failure of even the most well-intentioned acts may be unnoticed in the absence of these performance indicators. Not all performance indicators are helpful in all situations because of the variety of observational scales and themes. However, performance indicators must explain whether the condition changed for the better or worse and measure that change. The results of performance indicators are also more informative if they can be quantified to make comparisons across various activity kinds easier. However, targets and baselines must be outlined before choosing performance indicators. The chosen performance indicators must employ indicators and connect to the goals, trustworthy, reproducible, and capable of being produced in a time- and cost-effective way.

7.2 MAIN PERFORMANCE INDICATORS: HEADING AND LATERAL ERROR

The performance of an AAV can generally be defined using two main parameters, including heading error (ε_{head}) and lateral error (ε_{lat}). Also, other parameters, such as safety error, response delay, and efficiency, can be used to measure a robot's performance in detail. The ε_{head} is the difference between the vehicle's heading and the path's desired angle. The ε_{lat} is the lateral distance between the vehicle and the desired path. These two parameters must be defined when an AAV is developed (Figure 7.1(a)). The determination of whether the designed vehicle is accurate enough or not is based on these two parameters. In Japan, the required lateral error must be ≤8 cm because the average distance between crop rows is 66 cm, whereas the width of wheels or crawlers is ~50 cm. In this case, the maximum accuracy of 8 cm can be acceptable for AAVs (Figure 7.1(b)). Otherwise, the vehicle can damage the crop rows, and the efficiency of AAV can

DOI: 10.1201/9781003296898-7

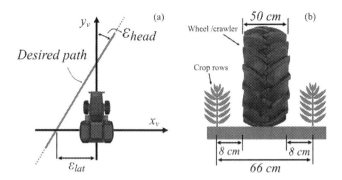

FIGURE 7.1 Performance indicators (a) ε_{head} and ε_{lat}, and (b) the maximum limit of lateral error.

be dramatically decreased. Based on the vastness of the land, the size of the agricultural field, the type of cultivation, the level of modernity of the agricultural process, and many other components involved in the cultivation, the acceptable range of these indicators can be different.

For the determination of the values of ε_{head} and ε_{lat}, the root means square error (RMSE) must be defined. Other indicators, such as accuracy, recall, precision, and determination coefficient, can also be used. The average error values cannot represent the performance because an AAV can have an average error of zero but still show high turbulence in lateral and heading control. The maximum and minimum values can be used as the sub-main parameters. After evaluating each AAV based on the indicators described in this section, the performance indicators of examined AAVs should be compared versus similar developments.

The main objective of the controlling algorithm is developing a kinematic and dynamic model to control the lateral and heading error of the vehicle and thus guide the vehicle along the desired path. The kinematic model of an autonomous vehicle must be defined as shown in Figure 7.2. The parameters include the coordination of the vehicle (x_c, y_c, θ_c), which indicated the position of the vehicle. The coordination can be measured by applying positioning sensors such as those used with RTK-GPS or any other position sensor, the reference coordination (x_0, y_0, θ_0), the objective position or desired location (x_d, y_d, θ_d), and the orientation of the vehicle (x_v) which is measured using attitude sensors such as an IMU, the steering angle (δ), the longitudinal velocity (V_x), the heading error (ε_{head}), and the lateral error (ε_{lat}). The general kinematic model for autonomous vehicles is shown in Equation (7.1). The dynamic model is shown in Equation (7.2).

$$\begin{bmatrix} \varepsilon_{long} \\ \varepsilon_{lat} \\ \varepsilon_{head} \end{bmatrix} = \begin{bmatrix} \cos\theta_c & \sin\theta_c & 0 \\ -\sin\theta_c & \cos\theta_c & 0 \\ 0 & 0 & 1 \end{bmatrix} \begin{bmatrix} x_d - x_c \\ y_d - y_c \\ \theta_d - \theta_c \end{bmatrix} \qquad (7.1)$$

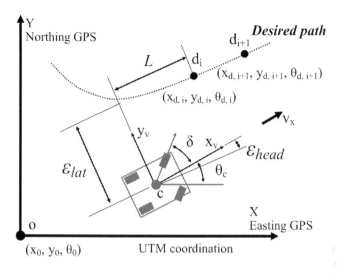

FIGURE 7.2 General kinematic model of AAVs.

$$\begin{bmatrix} \varepsilon_{long} \\ \varepsilon_{lat} \\ \varepsilon_{head} \end{bmatrix} = \begin{bmatrix} \cos\theta_c & 0 \\ \sin\theta_c & 0 \\ 0 & 1 \end{bmatrix} \begin{bmatrix} v \\ \dot\phi \end{bmatrix} \tag{7.2}$$

Then, the global coordination frame is shown in Equation (7.3):

$$\begin{bmatrix} x_c \\ y_c \\ \theta_c \end{bmatrix} = \begin{bmatrix} \int_0^t v \cos\left(\int_0^t \dot\phi \ dt\right) dt \\ \int_0^t v \sin\left(\int_0^t \dot\phi \ dt\right) dt \\ \int_0^t \dot\phi \ dt \end{bmatrix} + \begin{bmatrix} x_0 \\ y_0 \\ \theta_0 \end{bmatrix} \tag{7.3}$$

where ε_{long} is the longitudinal error and $\dot\phi$ is the yaw rate velocity. However, in the actual field conditions, the values of v and ϕ can vary due to the slip angle and different parameters related to soil features.

7.3 EVALUATION INDICATORS

Many algorithms were used to tackle various challenges in agricultural applications, requiring proper and reliable evaluation. This task can be performed using different evaluation indicators. The indicators compare the output of a model/algorithm with the experimental/actual data to determine the accuracy and performance of the model or algorithm. Several criteria were used to evaluate the algorithms: accuracy, precision, recall, the most common metrics followed by error-related metrics (i.e., MAE and RMSE), and the correlation coefficient.

7.3.1 ACCURACY

Accuracy has a positive correlation with the performance of ML methods and a negative correlation with RMSE (in general, with error-related metrics). Accuracy is the fraction of correctly classified samples among the total number of samples (Equation (7.4)):

$$Accuracy = \frac{True_p + True_n}{True_p + True_n + False_p + False_n} \tag{7.4}$$

where $True_p$ denotes the true positives, $True_n$, the true negatives, $False_p$, the false positive, and $False_n$, the false negatives.

Accuracy was used in different concepts as an analyzer and justifier for the output of a system. In research by Fatima et al., accuracy was employed for evaluating the output of machine learning to categorize multiclass fruits based on color. A better model based on hybrid color characteristics is created and provided to help the agriculture market. The outcome demonstrates that the accuracy was improved by 2% by combining the hybrid color feature with the linear kernel approach [125].

7.3.2 RECALL

Recall, also known as sensitivity, particularly in binary classification, is a metric that measures the relevance of a model. Equation (7.5) shows the formal definition of the recall metric, defined as the fraction of relevant instances of a class that are correctly classified (i.e., the class-conditional accuracy):

$$Recall = \frac{True_p}{True_p + False_n} \tag{7.5}$$

where $True_p$, denotes the true positives and $False_n$, the false negatives. In research, a recall was employed as a portion of outputs equal to the actual output values, especially in machine learning-based problems for detection, prediction, or estimation purposes. In this era, recall was employed by Tu et al. for comparing different concepts of convolutional neural network outputs for obtaining a proper detector of RGB-D images [126].

7.3.3 PRECISION

Precision is a metric that measures the overall performance stability of a model. Equation (7.6) outlines the formal definition of the precision metric, defined as the share of classifier decisions for a particular class that is correct:

$$Precision = \frac{True_p}{True_p + False_p} \qquad (7.6)$$

where $True_p$, denotes the true positives and $False_p$, the false positives. In the autonomous agricultural section, precision is frequently used for detection applications. This procedure plays a vital role in evaluating the performance of solutions for different problems and issues in the subset of precision agriculture. Huang et al. used a precision performance indicator to evaluate the convolutional neural network's detection performance in detecting tomato images in the subset of robotic vehicles for automatic tomato harvesting applications [127].

7.3.4 RMSE

RMSE (root mean square error) is an error-related metric that measures the difference between actual and predicted values. In general, increasing the difference between actual and predicted values reduces the accuracy and increases the error metrics such as the RMSE. Equation (7.7) defines the RMSE formally as follows:

$$RMSE = \sqrt{\frac{1}{N} \sum_{i=1}^{N} (x_i - \widehat{x_i})^2} \qquad (7.7)$$

where N denotes the total number of samples, x_i the actual samples, and $\hat{x}_i$, the predicted samples. RMSE can compare the output of a system with target and desired values to monitor the performance of the studied system. This indicator is employed in various applications and studies (for estimation, modeling, detection, and evaluation). Patel et al. employed different deep learning-based techniques for developing a proper algorithm for crop yield prediction on climatic constraints in the subset of precision agriculture [128].

7.3.5 DETERMINATION COEFFICIENT

The correlation coefficient measures the (linear) statistical relationship between actual and predicted values. In particular, a higher correlation between target and output values increases the overall accuracy and reduces total error. Equation (7.8) shows the formulation for calculating the correlation coefficient:

$$Determination \ \ Coefficient = \left(\frac{Cov(x, \hat{x})}{\sigma_x \sigma_{\hat{x}}} \right)^2 \qquad (7.8)$$

where x refers to actual samples, $\hat{x}$ to predicted samples, $Cov(x, \hat{x})$ to the covariance between x and $\hat{x}$, and σ to the standard deviation (calculated for both x and $\hat{x}$). The correlation coefficient ranges between -1 and $+1$. A negative number indicates a negative correlation, whereas a positive number denotes a direct correlation between target and output values: the closer the coefficient to 1, the higher the resulting correlation and the accuracy. Yang et al. evaluated the application of support vector regression in estimating grain moisture content by comparing the model's output with the target grain moisture content [129].

7.4   CONTROLLING LOGICS

Logic is the study of the methods and principles of reasoning, and reasoning means obtaining existing propositions and expressions. As autonomous navigation in agricultural fields has non-linear behavior, different logic or different control systems could be used to control lateral and longitudinal errors. A controlling logic is the backbone of autonomous control, and the final results are directly related to the logic implemented. In this between, the artificial neural network (ANN), fuzzy logic, a proportional-integral-derivative (PID) controller, and a model predictive control (MPC) play a significant role. Each logic has some specifications which make them suitable for different applications. The following sections explain each in detail.

7.4.1   ARTIFICIAL NEURAL NETWORKS (ANNS)

An artificial neural network (ANN) can be used to define the output of attitude sensors in non-linear conditions (Figure 7.3). When a neural network was used in 1992, GNSS and GPS were not sufficiently developed and/or too expensive for autonomous navigation. At that time, GDSs were commonly used as an attitude sensors. Neural networks were used to improve navigation performance because the image sensors (positioning sensors at that time) had a considerable interval and low accuracy. A neural network comprises three layers: an input layer, a hidden layer, and an output layer. The hidden layer could be a single layer or several layers

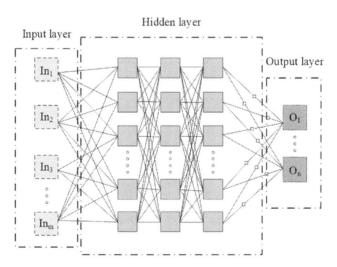

FIGURE 7.3   The architecture of an ANN.

based on the requirements. As the GDS was affected by the surrounding magnetic fields, the neural networks were modified to overcome the error of inclinations and magnetism. However, it was necessary to train the neural network, and the AAV had to be trained before the main maneuvers were performed. One of the significant disadvantages of neural networks was the black boxes, i.e., some unidentified boxes in programming due to the hidden layers. The presence of such boxes resulted in some uncontrollable behaviors in navigation that could make the AAVs dangerous.

An initial and straightforward approach to creating an intelligent learning system inspired by the biological neurons that make up brains is to use ANNs. Without the requirement to be programmed with task-specific rules, this system uses a training stage associated with a given task that pulls information from a training data set [130]. Performing tasks without a prior understanding of the nature of occurrences are the fundamental principle behind ANNs. As a result, ANNs can produce identifying traits (i.e., extracting discriminative features) from the input data [131]. Warren McCulloch et al. created the ANNs for the first time in 1943 [132]. This work simulated a primary neural network comprising electrical connections to examine how well neurons are and perform when doing learning tasks.

ANNs can be considered an all-encompassing modeling framework for handling complicated data sets. ANNs have been used for various applications, including curve-fitting, regression, and forecasting [130,133]. Neurons are the basic building blocks in an ANN model that use transfer functions to produce the output values. The main benefit of ANNs is that they are quick and inexpensive ways to deal with big data sets [134].

ANNs can be considered a comprehensive modeling framework for addressing challenging data sets. Curve-fitting, regression, and forecasting are only a few of the current uses for ANNs [135]. An ANN model's fundamental building blocks are neurons, which employ transfer functions to generate the output values. The primary advantage of ANNs is that they are quick and affordable methods for handling large data sets [136]. Figure 7.3 presents the structure of the ANN.

Figure 7.3 shows that the input weight matrix connects input and hidden layers. In contrast, the output weight matrix connects hidden layers and output layers in a straightforward feed-forward ANN. The model's training phase involves the learning of both matrices. To generate the output values, the ANN is therefore defined by the following equation (Equation (7.9)) [130]:

$$O = f\left(b + \sum_{i=1}^{n} w_i x_i\right) \tag{7.9}$$

where O is the output value from each node that will be affected by the bias b value, and w is referred to as weight values that regulate the propagation value x from input to output, with n being the number of layers. ANN has been widely used in the agricultural machinery sector for different applications. Table 7.1

TABLE 7.1
Notable ANN-Based Techniques for Agricultural Machinery Applications

Ref.	Year	Description	Analyzing Data Source	Evaluation Criteria	Application
[137]	2021	Crop yield prediction using ANN	Remote sensing data	R^2, RMSE	Estimation
[138]	2020	Paddy yield prediction	Climate data	R	Estimation
[139]	2020	Automatic detection of diseased leaves	Image processing	Accuracy	Detection
[140]	2020	Daily rainfall forecasting	Metrological data	R^2, RMSE	Estimation
[141]	2020	Olive moisture content prediction	Dielectric properties	R, MSE	Estimation
[142]	2016	Potato yield prediction	Remote sensing satellite	Accuracy	Estimation
[143]	2016	Agricultural crop prediction	Soil parameters	Accuracy	Estimation
[144]	2014	Monsoon rainfall forecasting	Metrological data	Accuracy	Estimation
[145]	2013	Fuel consumption prediction in wheat production	Farm and machine condition	MSE	Estimation
[146]	2012	Ripeness classification of oil palm	Color features	Accuracy	Classification

presents the notable cases that employed ANN in agricultural machinery. To train an ANN, different methodologies and algorithms, such as genetic algorithm (GA), imperialist competitive algorithm (ICA), particle swarm optimization (PSO), etc., can be used.

7.4.2 Fuzzy Logic

In classical logic, propositions are either true or false; that is, the value of a proposition is either zero or one. Fuzzy logic extends classical two-valued logic to propositions whose truth value can have any value in the range [0,1]. This generalization allows us to approximate reasoning. It means we can obtain imprecise and approximate results (fuzzy propositions) from a set of imprecise conditions (fuzzy propositions) [147,148].

Fuzzy logic has two different meanings. In the first sense, which has a more limited view, fuzzy logic is a logical system that results from multi-valued logic, but in a broader sense, it is almost synonymous with the theory of fuzzy sets. Fuzzy set theory is concerned with classes of objects with non-salient boundaries. In these classes, the membership of objects in each class is described by a concept called degree of membership. This article was published in America by Lotfizadeh, a professor of systems theory at the University of California, Berkeley. Since that date, fuzzy logic has followed various examples of recent technologies. When the concept of fuzzy logic was published in America, it faced many criticisms. The main problem taken from fuzzy logic at that time was that there was no practical way to implement it in the industry, which has been solved today [149].

The word *fuzzy* is defined in the Oxford dictionary as vague, dumb, imprecise, confused, confusing, confused, and unclear. Fuzzy systems are precisely defined systems, and fuzzy control is a particular type of nonlinear control that is also precisely defined. This article is similar to control and linear systems, where the word *linear* is a technical attribute that defines the state and condition of the system and control. The same is true of the word *fuzzy*. Although fuzzy systems describe non-deterministic and uncertain phenomena, the fuzzy theory is precise. In practical systems, important information originates from two sources. One source is experts who define their knowledge and awareness about the system or natural language. Another source is measurements and mathematical models that are derived from physical rules. Therefore, an important issue is the combination of these two types of information in the design of systems. The critical question is how to convert human knowledge into a mathematical formula to do the combination. What a fuzzy system does is transform [150,151].

In general, the reason for using fuzzy logic can be summarized in the following paragraphs:

- Fuzzy logic is conceptually straightforward because the concepts used in fuzzy reasoning are very simple.
- Fuzzy logic is very flexible. It means organizing a fuzzy system to solve a new problem, and there is no need to redesign the system.

- Fuzzy logic works based on the experience of experts.
- Fuzzy logic can be integrated with conventional control techniques. Fuzzy systems are not necessarily replacing conventional control systems; in many cases, fuzzy systems complete and facilitate the implementation of control systems.
- The principles of fuzzy logic are based on the type of human communication. Since fuzzy logic is based on qualitative description structures in everyday life, it is effortless to use.

Fuzzy systems are systems based on knowledge or rules. The heart of a fuzzy system is a knowledge base that consists of fuzzy if-then rules. A fuzzy if-then rule is an if-then statement, some words specified by continuous membership functions. A fuzzy system is a set of if rules. Then the fuzzy is made. In short, the starting point of building a fuzzy system is to obtain a set of fuzzy if-then rules from experts' knowledge or the field under investigation. The next step is to combine these rules into a single system. Different fuzzy systems use different principles and methods to combine these rules. The structure of a pure fuzzy system is shown in the figure below. A fuzzy rule base represents a set of fuzzy if-then rules. The fuzzy inference engine combines these rules into a mapping from the fuzzy sets in the input space to the fuzzy sets in the output space based on the principles of fuzzy logic. The main problem with pure fuzzy systems is that their inputs and outputs are fuzzy sets (words in natural language). While engineering systems, inputs and outputs are real-valued variables. A simple method is to add a fuzzifier to the input that transforms real-valued variables into a fuzzy set to use pure fuzzy systems in engineering systems. The result is a fuzzy system with a fuzzifier and a non-fuzzifier, shown in Figure 7.4.

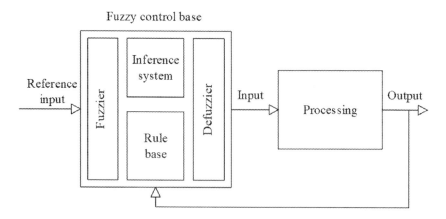

FIGURE 7.4 Fuzzy system structure.

The fuzzy controller has four main combinations:

1. The "rule-base" part keeps knowledge in the form of a set of rules that ultimately applies the best control to the system.
2. The inference mechanism evaluates which control rule is relevant, and then decides which input should be sent to the process.
3. The fuzzifier interface introduces the inputs so that they can be compared and interpreted with the rules in the basic rules.
4. The de-fuzzification mechanism turns the conclusion from the fuzzy inference mechanism into the process's input.

Systems based on fuzzy knowledge are among the most successful applications in fuzzy theory and fuzzy logic. With their simplicity and flexibility, fuzzy laws can express knowledge and theoretical developments in this field. Fuzzy modeling and control provide a framework for complex nonlinear relationships using a rule-based approach [152].

The heart of a fuzzy system is a knowledge base that consists of fuzzy if-then rules. A fuzzy if-then rule is an if-then statement whose continuous membership functions specify some words. A fuzzy system is made from a set of fuzzy if-then rules. In short, the starting point of building a fuzzy system is to obtain a set of fuzzy if-then rules from experts' knowledge or the field under investigation. The next step is to combine these rules into a single system. Different fuzzy systems combine these rules with different principles and methods [151,153].

Fuzzy logic is a many-valued logic. Three different functions handle the lateral error based on the steering angle to control a vehicle (Figure 7.5), defined as trapezoid-shaped curves. When the lateral error is almost zero, the f_0 controls the steering angle. When the lateral error starts to increase, the density of f_0 decreases and the system is going to use f_1. It is the same situation when the lateral error is decreasing. The functions can be developed based on the required parameters in the algorithm. Table 7.2 presents the notable fuzzy-based techniques in agricultural machinery applications.

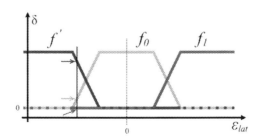

FIGURE 7.5   Fuzzy logic.

TABLE 7.2
Notable Fuzzy-Based Techniques for Agricultural Machinery Applications

Ref.	Year	Description	Analyzing Data Source	Evaluation Criteria	Application
[154]	2022	To monitor the manufacturer's innovations	A survey conducted with 103 Brazilian companies	NA	Identification
[155]	2022	To evaluate the effect of variation in agricultural machinery speed on the spray effect	Based on the 3WF-1000 sprayer	MAPE	Evaluation
[156]	2022	To study the digitalization and ecosystem-based capabilities of manufacturers	A systematic literature review	NA	Identification
[157]	2022	Crop pest estimation	Samples collected in a cropping cycle revealed the plausible correlation	NA	Prediction
[158]	2020	To improve the quality of harvesting	Data from the Hall sensor	NA	Decision making

7.4.3 PROPORTIONAL, INTEGRAL, AND DERIVATIVE (PID)

PID stands for proportional (P), integral (I), and derivative (D). The PID controller is universally accepted and commonly used in industrial applications because the PID controller is simple and provides good stability and fast response. Each application's coefficient of these three actions is different to obtain optimal response and control. The controller's input is the error signal, and the output is given to the device or process. The effort is to produce the controller's output signal so that the device's output reaches the desired value. The PID controller is a closed-loop system with a feedback control system that compares the process variable (feedback variable) with the set point, generates an error signal, and adjusts the system output accordingly. This process continues until this error reaches zero or the value of the process variable is set equal to the point (Figure 7.6) [159,160].

A PID controller is a type of controller that can correct the required navigation parameters using the feedback signal, which produces the error value. This controller uses three proportional-integral-derivative functions to control the error (Figure 7.7). PID controllers are now used as an appropriate control for autonomous navigation because of their high capacity and controllable functions.

Because of the disadvantages of neural networks and fuzzy logic, using a PID controller is becoming more popular nowadays. In this regard, mainly using two PID controllers, the dynamic model can be developed to control lateral (PID-1) and longitudinal (PID-2) motion by adjusting the steering angle and gas pedal command, respectively. The PID-1 controller is for lateral control, which consists of two control systems: a P control based on ε_{head} and a PID controller based on ε_{lat}, as shown in Equation (7.10) and Figure 7.6(a). The objective of this

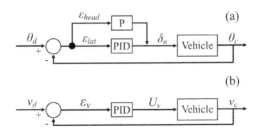

FIGURE 7.6 PID control system for (a) lateral control (PID-1) and (b) longitudinal control (PID-2).

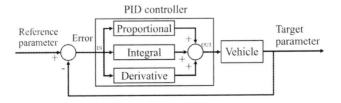

FIGURE 7.7 The theoretical principle of PID in an AAV.

control system is to control the vehicle's steering angle along the desired path. In this regard, the system tries to keep ε_{lat} and ε_{head} near zero.

$$\delta_n = K_{p,lat} \cdot \varepsilon_{lat} + K_i \cdot \sum_{i=1}^{n} \varepsilon_{lat}\Delta t + K_i \cdot \dot{\varepsilon}_{lat} + K_{p,head} \cdot \varepsilon_{head} \qquad (7.10)$$

The PID-2 controller is for longitudinal control (Figure 7.6(b)). This control system adjusts the current longitude velocity of the vehicle (v_c) based on the desired velocity (v_c). The velocity error (ε_v) comes to the PID-2 controller as an input value, and the PID-2 controls the gas command (U_v) of the ECU. It can be a pedal control output for the vehicle, controlled by external actuators. The control function is as follows:

$$U_v = K_p \cdot v_e + K_i \cdot \sum_{i=1}^{n} v_e\Delta t + K_i \cdot \dot{v}_e \qquad (7.11)$$

where

$$v_e = v_d - v_c \qquad (7.12)$$

For agriculture applications (Figure 7.2), simplified steering control is sometimes used as follows:

$$\delta_n = k_1 \cdot \varepsilon_{lat} + k_2 \cdot \varepsilon_{head} \qquad (7.13)$$

in which k_1 and k_2 are gains. The ε_{lat} error is:

$$\varepsilon_{lat} = \frac{\alpha E_{GPS} + \beta N_{GPS} + C}{\sqrt{\alpha^2 + \beta^2}} \qquad (7.14)$$

In which α, β, and C are constant coefficients in the linear equation. The E_{GPS} and N_{GPS} are RTK-GPS's easting and northing, respectively. The ε_{head} is:

$$\varepsilon_{head} = a\tan\left(\frac{\varepsilon_{lat}}{L}\right) + \theta_c \qquad (7.15)$$

Table 7.3 presents the notable PID-based techniques in agricultural machinery applications.

7.4.4 MODEL PREDICTIVE CONTROL (MPC)

Model predictive control (MPC) is an advanced control strategy first used in the oil refinery industry in 1981 to control the distillation column. In MPC, it is possible to

TABLE 7.3
Notable PID-Based Techniques for Agricultural Machinery Applications

Ref.	Year	Description	Analyzing Data Source	Evaluation Criteria	Application
[161]	2021	To reduce the great slippages from the complex conditions of field surface	The GNSS measurement data	MSE	Controlling
[162]	2020	To develop a hydraulic transplanting robot control system	Simulation data	Response time	Controlling
[163]	2018	To model the hydraulic system of the rice transplanter	Agricultural machinery and design criteria	Efficiency index	Controlling
[164]	2017	To control the tractor's front wheel angle	Real field test	Position deviation	Controlling

consider the system's physical constraints during the design. Another advantage of predictive model control is that this controller is the optimal control, but it is optimized in real time and online.

In recent years, predictive control has been widely used in many industrial processes and has provided good performance. Predictive control can pre-estimate behavior and take appropriate actions using linear or nonlinear models. Control signal changes are calculated by optimizing the specified objective function in a limited time horizon, and only the first control signal change is executed. The calculation is repeated for the following change. The feedback mechanism used makes the prediction errors caused by the mismatch between the model and the process to some extent be compensated.

The MPC is a feedback control algorithm that uses a model for a prediction about the feature of processes that can be used for predicting and optimizing an AAV, including steering angle and velocity. There are many reasons to use the MPC controller:

1. The MPC can handle multi-input and multi-output (MIMO) systems like ANN. This functionality is proper when it is predicted that there are some interactions between inputs and outputs (Figure 7.8). The MPC consists of three parts: prediction model, rolling optimization, and feedback adjustment. In an AAV, the steering wheel angle can affect the velocity output because of wheel slip or other possible reasons. The PID controller can complicate the controlling algorithm because two PID control loops operate independently. Designing larger systems will be challenging if there are no interactions between loops. The advantage of MPC is its multivariable controller that controls the outputs simultaneously by consideration of all the interactions between system variables.
2. The MPC can handle significant constraints because violating them leads AVs to undesired consequences. An AV must obey speed limits and maintain a safe distance from other vehicles, AVs (in the multi-robot system), and obstacles. They are also constraints due to AVs' physical limitations, such as limits on acceleration. If the AV works based on an MPC algorithm, the controller will track a desired trajectory while satisfying all these constraints.
3. The MPC has preview capability, similar to feedback control. When an AV travels on a curvy path or turns at the headland, if the controller

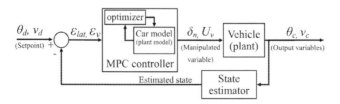

FIGURE 7.8   The theoretical principle of MPC in an AAV.

the first step. Therefore, if the mathematical model of the system is not accurate, the predictions of the system's output will not be valid and will lead to errors.

The complexity of solving the optimization problem for nonlinear systems is another disadvantage of predictive control. If the system's dynamics are non-linear, then the control cost function predicted by the model will become a complex function of decision variables (control signal during the control horizon), and its optimization will bring many problems. The nonlinear system can be converted into a linear system by defining a nonlinear transformation (mapping) to solve this problem. The linearizer feedback method is the same as in the nonlinear control theory. Table 7.4 presents the notable MPC-based techniques in agricultural applications.

8 External Attachments

8.1 INTRODUCTION

It might be complicated to comprehend end effectors. End effectors come in literally hundreds of varieties and come from several manufacturers. You could also hear the phrase "end of arm tooling" (EOAT). A robot's end effector is essentially its "business end," without which most robots are essentially worthless. Although an articulated robotic arm may be set to a specific place within its workspace, it is powerless to execute any operations without an end effector. The workspace of robotic manipulators is constrained. They might be mounted on portable robotic workstations to produce mobile manipulators and enlarge the workspace. Furthermore, mounting tiny robotic manipulators is advantageous since they frequently do not have a large payload to support a force sensor on the end-effector. The areas on mobile platforms where the manipulator may be attached are restricted [170].

8.2 ROBOTIC ARM AND MANIPULATION SYSTEMS

A robot manipulator is a multi-segment, electronically controlled machine that interacts with its surroundings to carry out tasks. They are also frequently known as robotic arms. In other words, a robotic arm is a machine designed to carry out one or more tasks quickly, accurately, and efficiently. These devices are often employed in industrialized countries today. Robotic manipulators have a wide range of specialized uses outside the industrial manufacturing industry. They are typically employed while carrying a heavy weight, doing repeated tasks, or working quickly. The arms typically comprise an arm, an end effector, controlling systems, and joints. The joint serves as the point of attachment, and it is here that the arms are joined. The end of the arm that manipulates materials or carries out various tasks is also referred to as the hand.

The placements and orientations of the many components that make up robot manipulators constitute an essential research topic. This module presents the fundamental ideas needed to characterize these rigid body locations and orientations in space and carry out coordinate transformations. Not only in the design of these arms, care should be taken, but their use should also be under the supervision of an expert engineer and in a standard working environment [171,172]. If all the stages of designing and implementing robotic arms are done correctly and adequately, work efficiency will increase to a great extent. In the 21st-century markets, speed is one of the factors that can quickly become your competitive advantage. Although the necessary condition to benefit from high

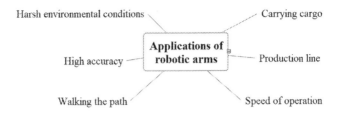

FIGURE 8.1 The different applications for robotic arms.

speed is good quality and market share, slowness will be detrimental to organizations [173,174]. Generally, any task that demands speed, accuracy, or repetitious actions can be delegated to a robotic arm. Robotic arms are being used on manufacturing lines in enterprises worldwide. Some employ a robotic arm only for a single surgery, while others use one for an entire series of surgeries. Figure 8.1 presents the different applications of robotic arms.

Table 8.1 categorizes each application for robotic arms in terms of the essential points, proper selection (advantages), and lack of proper selection (disadvantages). This table can quickly guide us in obtaining detailed insight into each application. If you have the right choice of robotic arms for the task at hand, these arms will work well and with great advantage. Otherwise, no particular advantage can be considered for that robotic arm, which will become a functional disadvantage (see Table 8.1).

TABLE 8.1
Performance Comparison of Robotic Arms in Agriculture

Operation	Important Application Points	Proper Selection	Lack of Proper Selection
Carrying cargo	Pay attention to the capacity	High profit and income	The risk of financial costs
Production line	Packaging the product, separating parts, or removing a part and placing it on another production line or terminal	High accuracy	Time-consuming process
Speed of operation	Personalized arms specific to the desired operation	High speed and accuracy	Time-consuming and high cost
Walking the path	For modern storage methods	High accuracy of movement	Shorter life span
High accuracy	Digital industries or watches	High accuracy	More expensive and more complicated planning
Harsh environmental condition	Environment exposed to toxic gases or very high temperature	High durability	More expensive

8.3   ROBOTIC ARMS

8.3.1   ARTICULATED ROBOTS

Similar to how the human arm attaches to the body at the shoulder, these robots have one arm and at least two joints. They were given such names because the robotic arm resembled a human arm. Each joint of this kind will provide the arm more mobility, regardless of the type's number of joints. An articulated robot has joints that are perpendicular and parallel to one another. Robots palletizing food, flat-glass handling, manufacturing steel bridges, cutting steel, heavy-duty robots, automation in the foundry industry, heat-resistant robots, and spot welding robots are examples of the application of articulated robots in different sectors (Figure 8.2) [175,176]. Examples of articulated robots are FANUC, Model: ArcMate 100iC (FANUC Co., Japan [177]), and ABB, Model: IRB 6600-225/2.55 (ABB Co., Switzerland [178].

8.3.2   CARTESIAN ROBOTS

They are also called rectangular robots. It is because of its limitation of the movement to coordinate axes in three dimensions. Cartesian robots can only move in the direction of three axes, X, Y, and Z. By adding an extra wrist, you can also provide them with the ability to move like a rotating movement. Cartesian robots can be controlled with single controllers. Cartesian robots are also known as multiple robots. A Farmbot Cartesian robot and Internet-based program design were built by Moscoso et al. for robot applications in farming and industrial facilities (Figure 8.3). Accordingly, electronic parts, firmware, robot work equipment, and artificial vision are implemented. Thirdly, no robots will collide during operation because of the programs' efficient design. Finally, the output of many apps is displayed [179]. Examples of cartesian robots are Festo, Model: YXCR (Festo Co., Germany [180]), and Yamaha Model: XY-XC (Yamaha Co., Japan [181]).

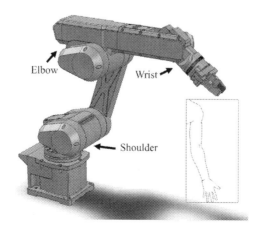

FIGURE 8.2   Articulated robots.

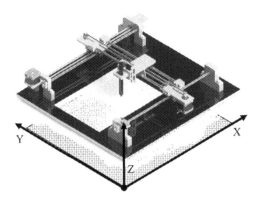

FIGURE 8.3 Cartesian robots.

8.3.3 SCARA Robots

Selective compliance assembly robot arm, or SCARA robots, consist of two parallel arms. This parallelism brings adaptability to the system in parallel work lines (doing the same work in two different work lines). The axes of rotation are vertical, and the place of impact of the robot (hand or wrist) is moving horizontally (Figure 8.4). The advantage of these robots over other robots, especially Cartesian robots, is a high speed and easy installation. The primary use of SCARA robots is related to assembling parts. An industrial and cost-effective SCARA robot's mechanical design approach was provided by Shariatee et al. Due to their innate

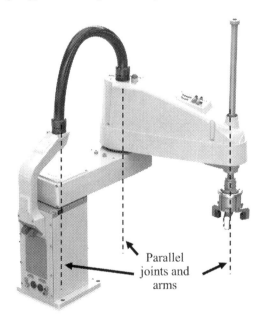

FIGURE 8.4 SCARA robot.

stiffness and great precision, these robots are among the most often employed in the industry. Utilizing locally accessible components while trying to keep prices down was a difficulty. This robot's control architecture makes it simple to build new control algorithms. Finally, a crucial route in the robot's workspace is traced using a PID controller. The results show low error during the rapid trajectory [182]. Examples of SCARA robots are Yamaha Model: YK-XG (Yamaha Co., Japan [183]), and FUNUC Model: SR-3iA (FANUC Co., Japan [184]).

8.3.4 DELTA ROBOTS

Another name for this category of robotic arms is parallel connection robots, consisting of two joints connected to a common base. This base is a simultaneous controller for both arms. The arms also control the robot's end effector (wrist or hand), so the operation is performed at high speed. Most of the popularity of this category is related to this speed. As a practical example, the assessments of serial-parallel Delta robots with complete orientation capabilities are the focus of Brinker et al. An energy-related dynamic model is built for four hypothetical extensions based on effective kinematic linkages. The effects of the extra serial chains on the actuation torques of the fundamental parallel Delta robot are examined by solving the inverse dynamic problem. Additionally, the advantages of each distinct approach are examined from a design standpoint and assessed using torque-related indices (such as energy consumption and root mean square torque) (Figure 8.5) [185]. Examples of Delta robots are FANUC Model: M-1iA/0.5S (FANUC Co., Japan [186]), and ABB Model: IRB 360 (ABB Co., Switzerland [187])).

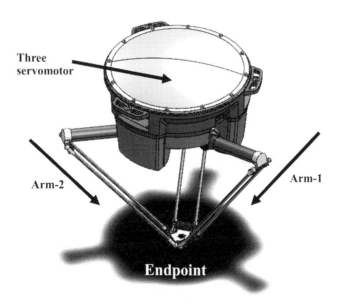

FIGURE 8.5   Delta robot.

8.3.5 CYLINDRICAL ROBOTS

The arm in this type of robot is connected to a cylindrical axis. These robotic arms can rotate around this cylindrical axis or vertically along the cylinder. A cylindrical robot can access tight spaces and perform operations without slowing down. In the study by Eden et al., using fruit location data gathered from 20 trees, a cylindrical robot was tested to describe the movements of a citrus-picking robot. The volume holding the fruit was divided into smaller volumes, and guessing the geodesic distance rather than computing it significantly reduced the computation time needed. However, in most situations, the TSP solution based on the estimated geodesic distance generated a fruit sequence quite similar to the one obtained using the precise geodesic distance between the fruit positions (Figure 8.6) [188]. An example of a cylindrical robot is PISHROBOT, Model: ARMC6MX28 (PISHR-OBOT Co., Iran [189].

8.3.6 POLAR ROBOTS

These are cylindrical robots that can rotate the arm in three dimensions. A movable arm is attached to the base with two swivel joints. This arm can move in 3D and perform various rotations. The accuracy of this robot is slightly higher than the cylindrical arm, but it still focuses on freedom of movement, and its main advantage is speed and range of motion. Another name for this category of robotic arms is "spherical robots" because their range of motion forms a sphere. Polar robots feature a sphere-shaped work area. Typically, a twisting joint and a mix of rotary and/or linear joints are used to unite the arm to the base. Spherical (twisting, rotary, and linear joint) or articulated (twisting, rotary, and rotary) robots are the two terms used to describe these machines; the latter more nearly mimics a human arm [190].

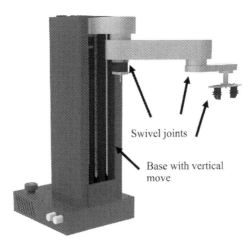

Swivel joints

Base with vertical move

FIGURE 8.6   Cylindrical robot.

TABLE 8.2

Advantages and Disadvantages of Each Application for Robotic Arms

Type	Application	Advantages	Disadvantages
Articulated	Food packaging, turning, welding, assembling the pieces, working with glass, casting	High speed, the most work coverage	Need a full-time guide, complex planning, and the need for very high mastery
Cartesian	Processes related to picking up and putting down objects, loading, and unloading, working with materials, assembling and disassembling, working with atomic materials	More flexibility, simple programming, and control, high movement accuracy	They are not small, need relatively large spaces, and movement restriction
SCARA	Put the pieces together, automatic packaging, moving the car, biological uses	High coordination and speed of repetitive tasks, perfect accuracy	Need a unique controller, and it is challenging to program, limited to the existence of a production line
Delta	The food industry, medicine industry, power industry, flight simulator, car simulator, optical fiber arrangement	Very high speed, very high accuracy	The need for a professional and full-time controller
Cylindrical	Moving LCD panels, coating, tinkering, casting, loading	Straightforward installation and setup, ability to work in tight spaces, occupying little space	Poor performance in the face of obstacles, low accuracy
Polar	Spot welding, working with casting machines, working with a gas welding furnace	More accuracy than the cylindrical arm, speed, and range of motion	High depreciation

Table 8.2 comprises different types of robotic arms in terms of the application type and the main advantages and disadvantages. Each type is suitable for a specific application and cannot be appropriately used in an application outside its capability.

8.4 END EFFECTORS

The agriculture industry has used robotic automation to meet the rising food needs. The primary agricultural chores that need time and work are fruit and vegetable harvesting. However, a seasonal labor scarcity of skilled employees causes low harvesting efficiency, food losses, and a decline in quality.

Research efforts are thus concentrated on automating manual harvesting activities. It is challenging to manipulate fragile objects with a robot in an unstructured environment. It is essential to build appropriate end effectors for manipulation [191].

Picking a fruit without destroying it or the crop is difficult once it has been identified and discovered. Fruits are typically hard to get at because so many unstructured barriers, such as branches and leaves, obstruct the system from collecting them. The speed and efficiency of a harvesting system must be balanced [191]. Vrochidou et al. categorized end effectors in agricultural applications into four main sets: contact-grasping grippers, rotation mechanisms, scissors/saw-like tools, and suction devices [191]. Figure 8.7 presents a proposed taxonomy for explaining the end effectors in agricultural tasks according to the categories presented.

One of the main reasons for using robots in agriculture is because the performance of repetitive and physically demanding tasks increases the danger of musculoskeletal diseases in workers and, in some situations, contamination with chemical preparations. It is necessary to deploy robotic tools for direct physical contact with and manipulation of objects in agricultural production to reduce production costs and improve the effectiveness of operations. A precise 3D specification of the positions of all the objects engaged in the interaction is necessary for weed removal, pruning and thinning, seeding, and fruit harvesting (Figure 8.7). The difficulty of the work arises from the picture

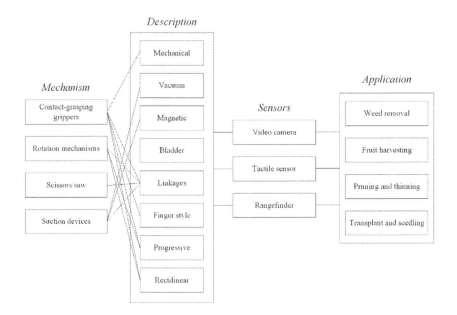

FIGURE 8.7   Different types of end effectors in agricultural applications.

analysis being done on a complex background with overlapping and similar objects (such as garden trees and numerous plants) (leaves, branches, and fruits) [192]. Since the physical and geometrical variations of fruits and the methods used to harvest them make it impossible to construct universal robotic systems, research is currently being done on designing mechanisms for processing fruits from various cultural traditions [192]. The main categories of end effectors used for agricultural tasks are listed in the proposed taxonomy.

Designing the end effectors is crucial because they are crucial to automated fruit harvesting [193,194]. End effectors have been created for different fruits and vegetables that utilize robots' capacity to distinguish between environmental variations associated with various crops [195–199]. These end effectors harvest fruits in three processes. They initially grab or suck the fruit. Second, they rotate, tug, or cut the fruit to separate it using water knives, lasers, or scissors. The fruit is then discharged into a collection box, and the end effector is reset. Most end effectors need three proximity processes: grabbing, picking, and unloading to harvest the fruit [200,204].

The success of complex control processes depends on identifying their strengths and weaknesses and extracting their accumulated errors. There is a need to compare and discuss the advantages and disadvantages of each mechanism to ensure the success of the tools related to agricultural operations to increase harvest efficiency. This section presents a simple comparison of end-effectors in terms of performance. Table 8.3 presents the main comparison categories. Table 8.3 also presents the advantages and disadvantages of each type.

While the human resources crisis is driving the need for automation in harvesting operations, research into the ground and aerial robotic harvesting systems is driven by the growing capacity for harvesting with high consistency and speed. Additionally, as older generations of farmers retire, adopting current technical

TABLE 8.3

A Comparison of Each End Effector Type for Agricultural Applications

Type	Performance Score	Product Damage	Advantages	Disadvantages
Contact-grasping grippers	Very high	Low	High accuracy	More cost and complexity
Rotation mechanisms	Median	High	Lightweight and flexibility	Complexity and low durability
Scissors/saw	High	Median	Simple mechanism and high durability	High risk and blade broken
Suction devices	High	Very high	High reliability	Low accuracy

FIGURE 8.8 Examples of end effectors for (pumpkin harvesting [203]).

developments in the agricultural industry may attract a new generation of farmers [191]. Huang et al. investigated the performance of an autonomous banana-picking robot with a rotational end-effector type (Figure 8.8) [201], thanks to a valuable review paper by Vrochidou and Tsakalidou [202].

9 Software and Apps

9.1 INTRODUCTION

In literature, applications (apps) are packages that carry out particular user tasks. Although all apps might fall within the software category, the opposite is not conceivable. Software is a group of apps that works together with hardware to operate a machine. A set of instructions or data controls the computer. Software is the polar opposite of hardware, the physical component, and it completes hardware in a computer. Apps and software have different applications. This section presents the taxonomy of software and app classifications (Figure 9.1).

9.2 SOFTWARE AND APPS FOR CONTROLLING

This section presents a general view of Myrobotlab, Microsoft Visual Studio, Processing, Notepad++, and ArduinoDroid.

9.2.1 MYROBOTLAB

Myrobotlab is a Java service-based open-source framework for robotics and creative machine control. It utilizes the Java 1.8 (or higher) JVM, and it could be used on any computer or device that supports this JVM. Numerous MyRobotLab Services will function on the Dalvik JVM for Android. The services offered by Myrobotlab include motor control, servo control, GUI control, voice recognition from Sphinx 4, text-to-speech from FreeTTS, and communication between microcontrollers. Myrobotlab Service wrappers encapsulate external services, which are then added to the framework [205].

9.2.2 MICROSOFT VISUAL STUDIO

In 1997, Visual Studio 97, with version number 5.0, became the first iteration of Visual Studio. Visual Studio's most recent version, 16.0, was made available. It also goes by the name Visual Studio 2019. Network framework's latest versions of Visual Studio range from 3.5 to 4.7. Older versions of Visual Studio support Java. However, the most recent version doesn't include any Java language support. Microsoft's integrated development environment (IDE) is called Visual Studio. It creates computer programs, websites, web applications, web services, and mobile applications. Microsoft's software development platforms, including Windows Store, Windows Presentation Foundation, Windows API, and Windows Forms, are used by Visual Studio. Both native and managed code can be generated by it.

DOI: 10.1201/9781003296898-9

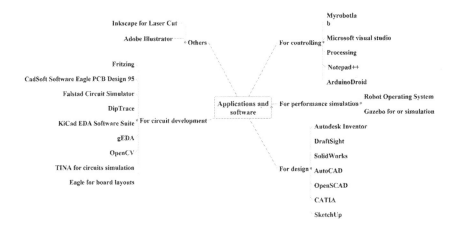

FIGURE 9.1   The taxonomy of the application and software for AAV development.

Developers can write and edit their code using Visual Studio. Code editing, debugging, and building are done through the software development's user interface. The code editor with Visual Studio supports code refactoring and IntelliSense, the code completion component. The integrated debugger performs both source-level and machine-level debugger performs both functions. A code profiler, a designer for creating GUI apps, a web designer, a class designer, and a designer for designing database schema, are additional built-in tools [206,207].

If a language-specific service is available, Visual Studio enables the code editor and debugger to support (to varying degrees) practically every programming language. It currently supports C, C++, C++/CLI, Visual Basic.NET, C#, F#, JavaScript, TypeScript, XML, XSLT, HTML, and CSS. Support for additional languages, including M, Node.js, Ruby, Python, and others, is possible through plug-ins. In the past, Java (and J#) were supported. The Community edition of Visual Studio is the free version and is the most basic version. "Free, fully-featured IDE for students, open-source and independent developers" is the tagline for Visual Studio Community version.

9.2.3 PROCESSING

Processing is a customized software sketchbook and programming language for the visual arts. Processing has fostered visual literacy in technology and software literacy within the visual arts since 2001. Processing is used for learning and prototyping by tens of thousands of students, artists, designers, researchers, and hobbyists. To teach non-programmers the foundations of computer programming in a visual context, the electronic arts, media technology art, and visual design industries created the free graphical library and IDE known as Processing [208].

Java is the language used by Processing, with additional simplifications like extra classes and aliased mathematical operations. Additionally, it offers a graphical user interface to make compilation and execution easier. Other projects like Arduino,

Wiring and Processing were all born from the Processing language and IDE. In reality, every Processing sketch is a subclass of the PApplet Java class, which implements the majority of the functionality of the Processing programming language. PApplet was once a subclass of Java's built-in Applet.

When writing code for Processing, all additional classes declared will be handled as inner classes before the code is compiled. It implies that static variables and methods cannot be used in classes unless Processing is expressly instructed to write in pure Java mode. Additionally, users in Processing can design their classes for the PApplet drawing. It circumvents the constraints of just utilizing traditional data types like int (integer), char (character), float (actual number), and color. Complex data types that can contain any number of parameters are now possible.

9.2.4 NOTEPAD++

Notepad++ is a multilingual source code editor and Notepad replacement that is free (both in the sense of "free speech" and "free beer"). The GNU General Public License controls how it is used when running in the MS Windows environment [209]. Notepad++ is developed in C++ and employs pure Win32 API and STL, guaranteeing a faster execution speed and reduced program size. It is based on the robust editing component Scintilla. Notepad++ aims to lower global carbon dioxide emissions by optimizing as many processes as possible without sacrificing user-friendliness. The PC may slow down and use less power while utilizing less CPU power, making the environment greener. Notepad++ is a source programming language. In contrast to intelligent code completion and syntax checking, it offers syntax highlighting, code folding, and limited auto-completion for programming, scripting, and markup languages. Although it may correctly display code written in a recognized schema, the validity of the syntax's underlying soundness or compilability cannot be confirmed.

9.2.5 ARDUINODROID

You can edit, create, and upload sketches to your Arduino board straight from an Android phone or tablet using the free ArduinoDroid software. The finest Arduino writing and building tool are Arduinodroid, which we suggest you use. The compiler supports all Android devices. The library is simple to use, compile, and transfer to the circuit board. Create your switch to operate the circuit board. Connect the USB OTG between the board and the phone; no computer is required [210].

Finding a compiler IDE that can compete with ArduinoDroid is quite tricky. Bring coding, compilation, and library running to a new level. Stop identifying and resolving bugs. This Arduino IDE program will look for a solution and automatically recommend one. Previously only feasible on the desktop IDE, this is now possible on your preferred Android phone. With a virtual keyboard that is easily accessible, learn an entirely new method to write CODE. Characters that are often used are suggested on the screen. Enter characters quickly without using the conventional keyboard's characters tab.

9.3   SOFTWARE AND APPS FOR PERFORMANCE SIMULATION (SIMULATORS)

This section presents a general view of the robot operating system (ROS) and Gazebo.

9.3.1   ROBOT OPERATING SYSTEM (ROS)

By integrating preexisting fixes for minor issues, developers may design an extensive system using the robot operating system, a framework for creating robot applications. It is an onboard computer and enables remote control and monitoring of a robot while running ROS on it in a virtual environment [211]. ROS is a collection of open-source robotics middleware. While not an operating system (OS), ROS is a collection of software frameworks for creating robot software. As such, it offers services like hardware abstraction, low-level device control, the implementation of frequently used functionality, message-passing between processes, and package management that are intended for a heterogeneous computer cluster. A graph architecture describes the running sets of ROS-based processes, with processing in nodes that may receive, publish, and multiplex messages relating to control, state, planning, actuators, and other topics. Although low latency and responsiveness are crucial for robot control, ROS is not a real-time operating system (RTOS). Real-time code can, however, be integrated with ROS [212,213].

The absence of real-time system functionality has been resolved with the development of ROS 2. A significant upgrade to the ROS API will support real-time code and embedded system hardware while utilizing contemporary libraries and technologies for fundamental ROS operations [214,215]. Due to their reliance on vast collections of open-source software dependencies, the primary ROS client libraries are designed with a Unix-like system. Ubuntu Linux is categorized as "Supported" for these client libraries, whereas other distributions, including Fedora Linux, macOS, and Microsoft Windows, are labeled as "experimental" and are supported by the community. However, these restrictions are not shared by the native Java ROS client library, ROSjava, which has made it possible to create ROS-based applications for the Android OS [216]. ROSjava has also enabled ROS to be included in a MATLAB toolbox that is officially supported and can be used with Linux, macOS, and Microsoft Windows [216].

9.3.2   GAZEBO

The gazebo is an open-source 3D robotics simulator. It combines the ODE physics engine with OpenGL rendering and supporting code for simulating sensors and controlling actuators [217]. ODE, Bullet, and other powerful physics engines are just a few of the ones that Gazebo can employ (the default is ODE). It renders realistic settings and includes excellent lighting, shadows, and textures. It may replicate sensors like wide-angle cameras, laser range finders, Kinectstyle sensors, and others that can "see" the simulated world [216]. The Gazebo uses the OGRE engine for 3D rendering.

With a complete toolkit of development libraries and cloud services, Gazebo takes a novel approach to simulation and makes it simple. Utilize high-quality sensor feeds and realistic settings to iterate quickly on your new physical ideas. Utilize simulation in continuous integration testing and test control methodologies for safety. A part of the Player Project from 2004 to 2011 was Gazebo. In 2011, Willow Garage began supporting Gazebo as an independent project. The Gazebo project was given to the Open Source Robotics Foundation (OSRF) in 2012 to oversee [218]; 2018 saw the rebranding of OSRF to Open Robotics.

9.4 SOFTWARE AND APPS FOR DESIGN (DESIGN AND SIMULATION)

This section presents a general view of Autodesk Inventor, DraftSight, SolidWorks, AutoCAD, OpenSCAD, Catia, and SketchUp.

9.4.1 AUTODESK INVENTOR

Autodesk Inventor is a computer-aided design tool for 3D mechanical design, simulation, visualization, and documentation [219]. The engineering design program Autodesk Inventor was created by Autodesk. In that both programs produce exact 2D and 3D models, Inventor and AutoCAD are both products of Autodesk. Tools for parametric, direct edit, and freeform modeling are available in Autodesk Inventor, along with multi-CAD translation capabilities and support for their standard DWG drawings. ShapeManager, a proprietary geometric modeling kernel from Autodesk, is used in Inventor. The direct competitors of Creo, SolidWorks, and Solid Edge are Autodesk Inventor [220].

When using Inventor for the first time, if you use AutoCAD or SolidWorks, the vocabulary and procedures are different, making the transfer challenging. In contrast to other 3D systems you might have used in the past, Autodesk Inventor is a parametric, feature-based system that enables you to build 3D components, assemblies, and 2D drawings. In comparison to AutoCAD's object-driven modeling, parametric modeling is quite different. Every aspect of a parametric modeler is governed by either parameters, dimensions, or relationships.

9.4.2 DRAFTSIGHT

For architects, engineers, construction service providers, professional CAD users, designers, educators, and enthusiasts, DraftSight is a feature-rich 2D and 3D CAD solution. This software features the capability, productivity tools, and file compatibility required to quickly and effectively create, edit, examine, and markup any 2D or 3D DWG file. Connecting to the 3DEXPERIENCE platform will improve design and documentation collaboration on the cloud even further. Figure 9.2 presents the advantages of DraftSight [221].

FIGURE 9.2   The advantages of the DraftSight.

9.4.3 SOLIDWORKS

SolidWorks is a solid modeler that builds models and assemblies using a parametric feature-based method that was first developed by PTC (Creo/Pro-Engineer). The program is created using the Parasolid kernel [222]. The term "parameters" refers to constraints whose values define the model or assembly's shape or geometry. Parameters might be geometrical terms like tangent, parallel, concentric, horizontal, or vertical, or numerical terms like line lengths or circle diameters. Numerical parameters can be connected by applying relations and capturing design intent. Design intent refers to how the part's creator wants it to react to updates and changes. For instance, regardless of the can's height or size, you would want the hole at the top to remain on the top surface. No matter what height the user later gives the can, SolidWorks will respect the user's design intent if they specify that the hole is a feature on the top surface.

Solidworks makes it simple to model objects and create two-dimensional maps. In fact, by using the view and map design environment, you may create your shape without having to deal with the calculations necessary for industrial drawing and extract the three directions of your shape. The SolidWorks program includes a wide range of facilities and tools for work, including welding, casting, molding, plastic injection, welding components, stress analysis, and modeling the behavior and resistance of the part under various loads. Compared to other CAD tools, this program is highly intuitive and easy to teach. The performance speed of this software is faster than that of other software.

This software can also interface with all machining and analytical programs, with more significant and faster performance. Equations can be written in models with this program between various parameters and sizes. Additionally, it allows users to communicate between sizes and equations in the Excel environment using Design Tables. With this software, it is possible to show the user exploded views of the design while also creating films and animations that show the exploded model being put together or taken apart for easier understanding.

9.4.4 AUTOCAD

If you're a student of architecture, electricity, mechanics, civil engineering, geography, urban planning, etc., you'll comprehend the meaning of AutoCAD better. Otherwise, you'll probably look for the AutoCAD program. Even in industries like designing green spaces, manufacturing, facilities, plumbing, tools, and facility planning. Learning the AutoCAD application typically brings back memories of your time as a student working on assignments for school. AutoCAD is software for drawing and designing accurate and professional engineering and industrial maps. This exciting software is one of the popular products of the American Autodesk company [223].

AutoCAD is one of the most influential and popular software for 2D and 3D designs and technical drawings. In actuality, the word CAD is an acronym for computer-aided design and appears in the name of this software. Of fact, CAD has been used in numerous other sources as a shorthand for computer-aided drafting, which refers to technical drawing with the aid of software. To put it simply, AutoCAD allows you to rapidly and accurately sketch objects on a computer rather than by hand or with a ruler.

This program includes many features, such as the ability to use the Text Command section, support for multiple lines, exclusive DWG extension, annotation capability adjacent to maps, high-quality processing and printing of designs, and file recovery. The user environment is very versatile and much simpler to use in this edition to complete complex and professional projects thanks to the ability to draw in both 2D and 3D, prevent undesirable changes, draw in true 3D, and use free design and drawing tools in both 3D and 2D settings. Additionally, it has an advanced programming environment for specialized applications. It can combine Excel and AutoCad-related components, complete management of map layers, and simple setting transfer between various computers, making it simpler to share designs between users with the least volume and highest quality.

9.4.5 OpenSCAD

OpenSCAD is a CAD program that is open source and enables users to produce 3D models and designs quickly. Unlike 3D design programs like Blender, OpenSCAD emphasizes CAD while producing 3D designs. Therefore, OpenSCAD will be more beneficial for you if you work with industrial designs, such as machine parts. OpenSCAD is not a program for interactive modeling. Instead, the application uses preset template files to produce 3D models (scripts). Users have complete control over the 3D modeling process and can alter any step of the modeling process at any time, thanks to this software structure. Users can therefore produce 3D models that are better suited to their preferences. AutoCAD DXF files, STL files, and OFF files may all be read by OpenSCAD to produce 3D models.

9.4.6 CATIA

Product life cycle management is a set of standards used by different sectors of the economy to regulate fundamental changes in product engineering and reduce production costs. Various stages of the product development cycle, such as concept design, product definition, manufacturing, simulation, and marketing expertise, use it as a starting point. Additionally, it offers general and geometric data pertinent to each stage of the production cycle [224]. CATIA was created by

Dassault Systemes, which is a multinational software company in France. CATIA software is commonly used in 3D design, computer-aided engineering solutions, PLM, and computer-aided manufacturing. This software is frequently used in the manufacturing and original equipment manufacturer (OEM) industries to expedite creating, assessing, and managing new products.

The user experience and product design are improved by CATIA software's usage of several points of view during the design stage. Additionally, the default tools for various components are enhanced throughout the product development process. As a result, it offers the greatest services to mechanical engineers, system architects, and creative and industrial designers. This 3D design platform allows collaboration in the modeling industry and social and online sharing of product concepts. It uses an interdisciplinary development platform and a diverse strategy. Thanks to the fantastic 3D user experience, practical workflow, and social design environment, it's effortless and straightforward. Advanced-level modeling structure and unbreakable relational design are the foundations of this program. This software allows users to simulate the behavior of products. Additionally, it aids in designing distributed, electronic, and electrical systems. In addition, compared to other 3D programs, its graphical user interface is more approachable, and the workspace's tools and instructions are apparent.

9.4.7 SketchUp

One of the best programs for creating 3D drawings is called SketchUp. This software is the simplest and quickest architectural 3D modeling program and is primarily used for envisioning and creating the basic shapes of designs. You may easily model the most exact volumes with numerous plug-ins [225]. Many users use the tools and capabilities of SketchUpsoftware for a variety of tasks, including developing and presenting booths, interior design, and industrial design. Designers and architects are highly interested in SketchUp because it allows them to focus on their designs and model their projects with only the most essential tools without having to deal with complicated settings or specialized software patterns.

Using the SketchUpsoftware, you may include many patterns and concepts you desire in the design into the AutoCAD designs. The ease of use, simplicity and compact size of the SketchUpsoftware are some of its essential qualities. Additionally, SketchUp has an exemplary user interface, and many users find it simple to use and pleasant in its surroundings. Users can make the designs they want and share them with other users using the appropriate 3D libraries in SketchUp.

9.5 SOFTWARE AND APPS FOR CIRCUIT DEVELOPMENT

This section presents a general view of Fritzing, CadSoft Software Eagle PCB Design 95, Falstad Circuit Simulator, DipTrace, KiCad EDA Software Suite, gEDA, OpenCV, TINA, and Eagle for board layouts.

9.5.1 FRITZING

Fritzing is an open-source and free program for electrical services. Its features include two-dimensional circuit assembly on the board, schematic creation, and printed circuit board (PCB) creation. This software is a tool that enables you to create circuit prototypes or diagrams before actually putting them together [226]. It was created specifically for people who need to build electronic projects, particularly those involving free hardware, yet lack the resources to do it. This tool also has fantastic community support that keeps it updated or is ready to assist you if you run into issues. It can also be a helpful resource for users who want to share and document their prototypes, for electronics professors and students, for professionals, and even for users in the general public. This program makes use of the Qt framework and is written in C++. All its source code is accessible through GitHub repositories, and broken up into several repositories for software and other components, such as Fritzing-App and Fritzing-Parts.

9.5.2 CADSOFT SOFTWARE EAGLE PCB DESIGN 95

CadSoft is a powerful EDA software owned by Autodesk company. Design automation means electronic design automation, and it can design and edit PCBs and schematic capture.

A central processor connects the board's fixed wires to other components in electrical circuits. One or more printed circuit boards can be found in almost all electronic devices. For more than 20 years, the popular CadSoft EAGLE software has served as a flexible and robust PCB design tool that can match the high-level capability of pricey commercial circuit board design software at a fraction of the price. EAGLE is also very easy to use, learn, and purchase. The function of DesignLink is an automatic connection to Premier Farnell's database that allows users to search for and locate parts without ever leaving the CadSoft EAGLE design environment. With only one mouse click, customers may identify design faults and request prototypes from top manufacturers using CadSoft EAGLE's ground-breaking PCB Services interface.

9.5.3 FALSTAD CIRCUIT SIMULATOR

Falstad circuit simulator is excellent Java-based software for electronic circuit simulators, testing new ideas, or troubleshooting a design prototype. It offers improved precision, excellent visualization tools, the capacity to save circuits for subsequent study, and the capability to add our components. Even for simple circuits, the additional procedures necessary to conduct a modest simulation might be scary because they involve an installation and have a higher learning curve [227]. When the software starts, you can see a schematic of a simple LRC circuit.

9.5.4 DIPTRACE

This software is a sophisticated EDA/CAD program that creates printed circuit boards and schematic diagrams for electronic and electrical engineers. With the aid of this tool, you may create any board, from the most basic to the most complicated, and use it in practical situations. The user also sees a 3D view of the final design at the same time as the printed circuit board design so that he may see a tangible preview and genuinely comprehend the outcome of the board. The DipTrace program is explicitly designed for engineers that operate efficiently and professionally in printed circuit board design and require a complete solution in this area. Among engineers and printed circuit board designers, this program is regarded as a specialized tool.

This software allows you to move models directly from the schematic design stage to the production stage. The design of circuits is accelerated by this software's sophisticated 3D tools and a collection of 3D models. This software's numerous import and export capabilities have increased its popularity. It is also entirely compatible with other CAD software. Beginners and those with little experience in PCB design can efficiently work with and learn this software because of its user-friendly interface and high learning capacity.

9.5.5   KiCad EDA Software Suite

KiCad is an open-source PCB design program with all the tools required to create a typical PCB. This software is quite user-friendly and comparable to other similar programs. It features schematic, PCB, and 3D environments. The KiCad software's schematic environment is equipped with all the tools to design electronic circuit schematics. Like other printed circuit fiber design software, using KiCad's schematic environment requires first bringing the necessary parts from the software library to the schematic environment. After that, you can link the components together and modify their data (such as quantity, part number, etc.).

9.5.6   gEDA

The gEDA is a software environment for electronic design automation and solving the problems mentioned. Users can utilize gEDA to design electrical diagrams, printed circuits, and simulations.

The printed circuit board, sometimes known as the PCB, is one of the most crucial components of any electronic project. A printed circuit board can be designed in three different ways. The first technique creates the project's schematic using specialized software, and the PCB is automatically converted. The bases that need to be connected are joined using a thoroughly mixed-up and busy tool in the second technique, and all these lines are then organized by pressing a key. In the third approach, there is no automatic sorting, and all the components must be placed on the fiber and connected with care and attention using a tool for drawing copper lines.

9.5.7   OpenCV

The task of conducting numerous calculations on images is one of the challenges faced by developers of visual software. Their processing requires the best

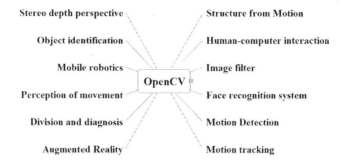

Stereo depth perspective   Structure from Motion

Object identification   Human-computer interaction

Mobile robotics   Image filter

OpenCV

Perception of movement   Face recognition system

Division and diagnosis   Motion Detection

Augmented Reality   Motion tracking

FIGURE 9.3   The applications of the OpenCV.

programming due to the volume of information. The Intel firm launched the OpenCV project to give developers of this type of software efficient libraries for real-time visual calculations [228]. OpenCV, or Open Computer Vision Library, is a set of image processing and machine learning programming libraries. Real-time image processing is more of an emphasis in this series. It was once created and sponsored by Intel, and now Willow Garage and Itseez provide support. The FreeBSD license makes use of it free of charge. Operating platforms such as Windows, Linux, macOS, iOS, and Android all support the cross-platform library known as OpenCV. Additionally, it features a programming interface for the languages C, C++, Python, Java, and MATLAB. Figure 9.3 presents the applications of the OpenCV software.

The basis of OpenCV's programming interface is written in C++, as is the application itself. Additionally, it offers a complete programming interface for MATLAB/Octave, Java, and Python. These languages' APIs are listed in the online documentation for OpenCV. Additionally, wrappers for programming languages like C#, Ch, and Ruby were developed to promote acceptance by a larger audience.

9.5.8   TINA

TINA for circuit simulation is a complete software package for the real-time design, analysis, and testing of analog, digital VHDL, MCU, and electrical circuits. DesignSoft Tina Design Suite is an engineering tool for circuit simulation and PCB design. The DesignSoft Tina Create Suite program provides a complete electronic laboratory. It allows users to design and simulate the electronic circuits they require, including analog, digital, symbolic, RF, and mixed-mode modeling and circuit design. The DesignSoft Tina Design Suite software is incredibly sophisticated and includes fast multi-core engines [229].

The features of DesignSoft Tina Design Suite include:

1. Analyzing, designing, and testing analog, digital, telecommunication, optoelectronic, RF, and VHDL circuits

2. It has an oscilloscope, function generator, multimeter, signal analyzer/bode plotter
3. It has a summary of data sheets of applied ACs
4. Educational files in the field of filter design, VHDL
5. PCB design
6. 3D viewing of circuit parts and 3D viewing of the circuit board design on the PCB

9.5.9 EAGLE FOR BOARD LAYOUTS

Engineers in electronics, medical engineering, etc., must acquire software for creating a printed circuit or PCB due to the integration of electronics into many engineering sciences fields. They must master PCB design software to create an electronic device. EAGLE software is the best, fastest, and most useful software in this field if you don't have enough time to study complicated and time-consuming PCB design tools [230]. EAGLE's software is cross-platform. Any platform can use and implement it, including Linux, Mac, and Windows. Not all PCB design program has this particular function. Despite having good quality and resolution and being similar to other software now on the market, EAGLE takes up a substantially lower amount of space.

EAGLE undoubtedly has issues, just like any other piece of software. For PCB design, some software is more effective than EAGLE, where features like the simulator, 3D vision, programmer, etc., are more effective. But in our opinion, EAGLE software is the most acceptable option for creating basic, intermediate, and complex circuits. Additionally, EAGLE offers the most pleasing environment for you to experiment and learn if you have never developed a printed circuit board and are a beginner in this sector.

Easy Applicable Graphical Layout Editor, or EAGLE software, is a fully functioning and user-friendly program with an extensive library that can be quickly learned and utilized in the design of electrical projects. Fusion 360 software now allows for the 3D design of circuits and products after Autodesk's acquisition in 2016. In this course, you will learn about circuit schematic design, PCB design, how to make a new part, how to get the output, and how to view the board in 3D using the EAGLE software.

10 Review of a Case Study

10.1 INTRODUCTION

This chapter is an abstract of a paper presented by Roshanianfard and Noguchi [231] about autonomous agricultural vehicles for arable crops developed in the laboratory of vehicle robotics (VeBots), at Hokkaido University between 1990 and 2018 [232]. All inquiries made in this laboratory are analyzed and assessed as an operational case study. The development process, component selection, developmental obstacles, and vehicle performance indicators are covered in detail. Following the presentation of each component's development and applicability, suggestions for additional research are made. The most significant events and lessons are listed, along with some suggestions.

These robots have undergone several evolutions, from a path planning system in 1997 to a multi-robot tractor (multi-RTs) system and intelligent systems [233] in 2018. Twelve projects are introduced in this section with 'AAV-n' titles. The n is the project number. Table 10.1 provides a general description of each project and summarizes the various platform types, transportation systems, and mobile platforms. The development procedure for autonomous agriculture vehicles can be divided into three eras: the development of technology (AAV-1 to AAV-3), commercialization (AAV-4 to AAV-11), and intelligent systems (AAV-12 and AAV-13). During the development era, the main target was designing, manufacturing, and evaluating an autonomous tractor that could maneuver in fields with little human interposition. At the end of that approximately 8-year era, the AV-3 could maneuver autonomously with high accuracy.

After this technology was established, laboratories decided to commercialize their systems and apply them on different platforms. A collaboration between Hokkaido University and the University of Illinois produced the AAV-4. The VeBots laboratory then developed another system by applying an autonomous system to a crawler-type tractor (AAV-5; model CT801, Yanmar), a utility vehicle (AV-6; E-Gator, John Deere), a commercialized wheel-type tractor (AAV-7; model EG83, Yanmar), a combine harvester (AAV-8; model AG1100, Yanmar), an airboat (AAV-9; model RB-26, Yanmar), a rice transplanter (AAV-10; model EP8D, Kubota), and a half-crawler-type tractor (AAV-11; model EG105, Yanmar). After successful technology expand, the multi-robot tractor (AAV-12; model EG453, Yanmar) and the robotic harvesting system for heavy-weight crops (AAV-13; model YT5113, Yanmar) developed. Detailed explanations of each system are presented in the following sections.

TABLE 10.1

General Description of AAV Projects at Hokkaido University

	Project No.	Type of Platform	Transportation System	Brand	Model	Project Name	Start	Finish
Development of technology	AAV-1[1]	Tractor	Wheel-type	Prototype	Prototype	NA	1992	1997
	AAV-2	Tractor	Wheel-type	Kubota	GL320	BRAIN[2]	1993	1998
	AAV-3	Tractor	Wheel-type	Kubota	MD77	NEDO[3]	1997	2000
	AAV-4	Tractor	Wheel-type	Case-IH	Magnum 7220	UIUC[4]	1997	1999
	AAV-5	Tractor	Crawler-type	Yanmar	CT801	Yanmar co.	2002	2008
Commercialization	AAV-6	Utility vehicle	Wheel-type	John Deer	E-Gator	Yanmar co.	2005	2008
	AAV-7	Tractor	Wheel-type	Yanmar	EG83	MAFF[5]	2010	2014
	AAV-8	Combine harvester	Combine harvester	Yanmar	AG1100	MAFF	2010	2014
	AAV-9	Airboat	Propeller	Yanmar	RB-26	NA	2012	2016
	AAV-10	Rice transplanter	Wheel-type	Kubota	EP8D	BRAIN	2014	2018
	AAV-11	Tractor	Half-crawler	Yanmar	EG105	SIP[6]	2015	now
Intelligent systems	AAV-12	Tractor	Half-crawler	Yanmar	EG453	SIP	2015	now
	AAV-13	Tractor	Half-crawler	Yanmar	YT5113	SIP	2016	now

Notes

1 AAV: Autonomous Agricultural Vehicle

2 BRAIN: Bio-oriented Technology Research Advancement Institution

3 NEDO: New Energy and Industrial Technology Development Organization

4 UIUC: University of Illinois Urbana-Champaign

5 MAFF: Ministry of Agriculture, Forestry, and Fisheries

6 SIP: Cross-ministerial Strategic Innovation Promotion Program

10.2   THE DEVELOPMENT PROCESS OF AAVS

Noguchi and Ishii [94] introduced the first robot tractor from the VeBots lab in a paper titled "A Study of an Intelligent Industrial Vehicle with a Neural Network" at the Second Intelligent Systems Symposium—in Japan in 1992. It was the first step in developing an intelligent autonomous vehicle for agricultural applications. Before that study, several research groups had designed agricultural vehicles controlled on a predetermined path using mechanical, optical, ultrasonic, or radio guidance and, in some cases, using leader cables. Each of these methodologies has some disadvantages (e.g., significant errors, lack of communication, and positioning errors), making them unsuitable for use with an auto-guidance system. At that time, the lack of a stable and accurate position-sensing system necessitated the development of different positioning systems.

The GNSS and GPS became available in 1990 only for military applications, and Noguchi and Ishii [234] then developed a mobile positioning system [42] and developed a mobile agricultural robot that used a geomagnetic direction sensor (GDS) and image sensors [235]. Their objective was to establish a mobile robotic system for agricultural applications that can detect its location using a positioning system and a heading angle sensor. They investigated a mobile robot that could be controlled by positions obtained from the image sensors and heading angles from a GDS. As shown in Figure 10.1 (AV-1), this mobile robot tractor (AAV-1) was a small prototype, rear-wheel-drive tractor with a petrol engine that could electrically control the steering angle (ϕ) using a potentiometer (max. $\pm35°$ that improved to $\pm40°$ [236]) and the rotation of a rear wheel by a rotary encoder (max. $7°/s$). It was also equipped with a controllable clutch and a brake using a computer which provided movement speed in the range of 0.4–1.2 m/s. In this AAV, the image sensors were used for the positioning system.

In a robotic system, not only the location of the robot should be predetermined as auto-guidance input, but the system's orientation should also be recognized. In 1996, GDSs were popular as low-cost sensors for measuring the heading angle (ψ) and mounted GDS almost 1 m above a robot tractor to avoid the noise effect. They also corrected the GDS outputs using an inclination sensor. The heading angle could measure roll and pitch directions in the range of $360°\pm6°$. Using a GDS involved two main errors: the error due to the magnetism around the robot and the error from the robot's inclinations. In this regard, Noguchi and Ishii [234] developed a neural network to refine the GDS output. Their neural network was composed of two networks: a correction network for vehicle inclinations and a correction network for the magnetism surrounding the robot. The neural network training data were obtained by giving the tested robot the inclinations in the outdoor environment.

Ishii and Terao [237] compared their neural network and a PID controller to identify the best control system for a robot tractor as a vehicle with a nonlinear kinematic system. The results indicated that the neural network controller, with a maximum error of 7.5 cm and 0.105 rad, had an accuracy higher than that of the PID controller (max. error of 31 cm and 0.195 rad). Noguchi and Terao [236]

FIGURE 10.1 The configurations of the autonomous vehicles designed at Hokkaido University [232].

later developed a control technique that used a combination of a neural network and a genetic algorithm (GA) that was able to create a suboptimal path for the developed robot tractor. Their research showed that the RMSE of forward, lateral, and yaw angular velocities were 0.0171 m/s, 0.0155 m/s, and 0.0156 rad/s, respectively. The developed method effectively determined a maneuvering path for the robot tractor.

Based on the control algorithm and guidance geometry, the robot could measure the baseline angle using the GDS before the robot's autonomous operation. Next, the target positions (previously measured using the positioning system) and the order of positions for arrival are indicated to the robot. After this procedure, the robot turns to the direction of the first target position using the data from the GDS. The straight-line path can be created from the current position to the first target position. A trial run of the robot was conducted on grassland at 0.5 m/s [238], and the results showed that the RMSE of the neural network method (1°) was almost

20% that of the conventional method (5.7°). The absolute maximum error and the RMSE of the position for the predetermined path were 51 cm and 23 cm, respectively. The maximum direction errors for the conventional and neural network methods were approx. 14° and 1°, respectively. The average error of the final position for each target position was found to be approximately 40 cm. This error was caused by measurement error of the angles, a time delay, and the measurement interval of the positioning system.

The results indicated that the neural network method was more effective than the conventional method because the angle detected from the GDS indicates more precise values. This result can be attributed not only to the position error but also to the poor steering response due to the lack of power of the stepper motor with which the robot was equipped. It seemed that the position error of the robot was satisfactory for transport work in a field. Still, the measurable area of the positioning system should be increased for practical uses. In addition, more accurate measurements of the robot's position are required for various farm operations.

Noguchi and his colleagues successfully developed their first autonomous vehicle, but the navigation system and interval of the vehicle required some modification. The accuracy of a positioning system should be optimized, and the communication time delay between a mobile vehicle and a station should be kept to a minimum. Yukumoto and Matsuo [239] developed an autonomous agricultural robot that can find a position with an error <5 cm [240]; the positioning error was almost 40 cm when they used image sensors and the principle of triangulation. In another study, Yukumoto and Matsuo [239] used a Kubota tractor (model GL320) [241] as a platform equipped with a TMS (geomagnetic) heading sensor, a control box, and a reflector for the navigation system. The configuration of this robot tractor (AAV-2) is named ROBOTRA.

For newly developed autonomous systems with limited experimental data, it is essential to evaluate different auto-guidance navigation systems. Yukumoto and Matsuo [239] designed and evaluated three different navigation systems: (1) a system using an off-the-wire method, named LNAV, (2) a method based on a combination of a DGPS and an inertial navigation system (INS) named "SNAV," and (3) an optical method named XNAV that uses two different types of reference stations, i.e., XNAV and AP-L1. LNAV was developed by Kubota Co. (Tokyo). SNAV was developed by Japan Aviation Electronics Industries (Tokyo). XNAV was designed by BRAIN-IAM (The Institute of Agricultural Machinery, Saitama, Japan), and XNAV by AP-L1 was developed by Topcon Co. (Tokyo).

The experiments were performed in a 100×150 m paddy field at 0.5 m/s. The results indicated the following positioning accuracy and interval values: LNAV, 50 mm and 0.1 s; SNAV, 100 mm and 0.1 s; XNAV, 50 mm and 0.52 s; and XNAV by AP-L1 10 mm and 0.5 s. However, the interval of the SNAV system improved when the system was equipped with an IMU. The XNAV results with the reference station AP-L1 were the most satisfactory.

The four navigation systems described above were costly, a commercialization limitation. To reduce the cost of orientation detection, Mizushima, Noguchi [242] developed a sensor fusion method (Kalman filter) that uses a GDS and a fiber optical gyroscope (FOG) based on dead reckoning. They constructed the kinematic model of the vehicle and conducted experiments with three autonomous guidance systems: GDS (FGM-300A) alone, FOG (JG35FD) alone, and the sensor fusion mode. They set three GDS or FOG sensors to make an IMU for measuring the roll, pitch, and yaw angles. The results showed that the lateral offsets of the GDS, FOG, and sensor fusion were 200, 400, and 61 mm, respectively, indicating that the sensor fusion can support maneuvers more accurately. They then applied the sensor fusion method on a Kubota GL320 platform, operated the vehicle manually, and collected the positioning data using RTK-GPS (MS750, Trimble Ltd.) with an accuracy of ± 2 cm\pm^2. The results indicated that the RMSE of the lateral offset was 38 mm, the maximum error from the regression line was 100 mm, and the angular error was 0.004°.

Mizushima and Noguchi [242] also reported that when the geomagnetic field around the sensor increased, the RMSE of the lateral offset, the maximum error from the regression line, and the angular error could increase to 92 mm, 170 mm, and 0.057°, respectively. However, their advanced sensor fusion method improved the auto-guidance performance more accurately than a single navigation sensor. The limitation of GDS due to the geomagnetic field could decrease the performance accuracy.

Mizushima and Ishii [243] later developed a low-cost attitude sensor composed of three vibratory gyroscopes (Gyrostar ENV-05F-03, Murata Manufacturing Co., Nagaokakyo, Japan), two inclinometers (D5R-L02-15, Omron Co., Kyoto, Japan), and a sensor fusion algorithm. This attitude sensor could measure the roll, pitch, and yaw angles with RMSE values of 0.43°, 0.61°, and 0.64°, respectively, to improve the positioning accuracy of the vehicle in a sloping field from 25.9 cm to 3 cm and in a bumpy field from 8.4 cm to 3.7 cm. This low-cost attitude sensor provided high accuracy at a price (USD 500) that was only 5% of the cost of previous IMUs, such as combined TMS or FOG sensors. To reduce the price of attitude sensors such as FOG and GDS, Liu, Noguchi [244] evaluated a low-cost IMU (S4E5A0A0, Seiko Epson Co., Suwa, Japan) using a mean filter for preprocession and Kalman filter sensor fusion. They aimed to provide acceptable accuracy and attitude.

The navigation systems dependent on satellite signals have shown weak performance in covered areas such as tunnels or orchards surrounded by trees. Since some agricultural fields include buildings, trees, or objects, sometimes RTK-GPS cannot work efficiently. Tsubota and Noguchi [245] thus developed an automatic guidance system that uses a 2D laser scanner (LM291) attached to the front of a robot tractor, a virtual reference station for RTK-GPS, and an IMU. Their test results showed that the lateral and heading errors of the system were 60 mm and 3°, respectively. These results were good enough to operate the robot tractor autonomously inside a coconut field [246]. Those authors recommended further modifications to improve this robot tractor's applications, capability, and performance by including different applications or the cooperation of several robot tractors.

To improve the efficiency of their developed autonomous tractor, Kise and Noguchi [247] designed another robot tractor to maneuver in a field based on a geographic information system (GIS) predetermined path for different applications. They evaluated a robot tractor engaged in different agricultural work in actual fields. The configuration of this robot tractor (AAV-3). It was a modified Kubota MD77 robot tractor [241] equipped with RTK-GPS (#MS750, Trimble, Sunnyvale, CA), FOG sensor (#JS35-FD, JAEI), and emergency switches. The data communication between the ECU and a personal computer was accomplished with CAN-bus. Because of FOG drift, a Kalman filter was applied to the control algorithm. This sensor fusion method was proposed by Michio et al. [56], and it could estimate the heading angle with 45-mm accuracy on different types of paths (straight, curved, and turns).

The AV-3 robot tractor was tested in a field at the operation speed of 1.5 m/s. The results indicated that the robot tractor could maneuver based on the predetermined path with a lateral error of 6 cm and a heading error of $1.3°$. The system could also implement rotary tillage with satisfactory parameters [248]. Noguchi and F. Reid [249] tested the same robot tractor by adding an IMU (a three-axis FOG sensor, #JCS7401A, JAEI) to the control configuration. They also used the same sensor fusion method and tested the robot tractor at a speed of 2.5 m/s in a soybean field in different operations, including tillage, planting, fertilizing, and cultivating [250]. The results demonstrated that the applied sensor fusion system with an IMU improved the accuracy of maneuvers, although the lateral error of the robot tractor was 30 mm. Those authors did not mention the heading error results in that study, but this error can be expected to improve [251]. Mizushima and Noguchi [252] obtained the same heading errors ($1.59°$, $0.41°$, and $0.65°$ in the yaw, roll, and pitch angles, respectively) when they used a microcomputer (#H8S2612, Renesas Co.) instead of a personal computer.

Because a robot tractor should be able to maneuver for various types of operation, the turning task at the end of a field remained a severe problem. Noguchi and Reid [253] addressed this problem by developing a switch-back turning function based on a spline function. They conducted this research through a computer simulation and real-world experimentation. The computer simulation determined the desired motion path and applied the developed function to the robot tractor (AAV-3). The turning function was composed of a forward and backward movement. The experimental results confirmed that the system could complete all turns successfully [254]. The positioning error was 70 mm overall for the turns [255].

Yokota and Mizushima [256] later installed a 2D laser scanner (#LMS291, SICK, Minneapolis, MN) on top of the robot tractor to generate a 3D GIS map that could be used for the collection of field information for precise farming. They could generate a map with a positioning error of 65 cm and an attitude error of 19 cm. Barawid and Oscar [61] installed the same laser scanner on the front of a tractor (70 cm above the ground) to develop an autonomous navigation system that can travel between trees in an orchard. After calibration by Barawid and Ishii [257], the tests at various speeds (0.36, 0.48, 0.62, 0.86, 1.23, and 1.43 m/s) indicated the best heading and lateral were $0.5°$ and 50 mm, respectively, when the robot's speed was 0.36 m/s. Those results, in comparison with the system developed by Tsubota and Noguchi [245] for the second robot tractor (AAV-2), revealed that the heading error and lateral error were improved from 60 to 50 mm and from 3° to 1.5°, respectively, demonstrating that autonomous navigation using different navigation and attitude sensors was significantly improved.

When the third generation of robot tractors was being developed, it was necessary to devise a communication system between two robots as multi-robot cooperation, a system that could connect two independent robot tractors with sufficient safety. Noguchi and Will [258] presented a master-slave system for farm applications. Their system consisted of two algorithms: a GoTo algorithm (the slave is guided to a predetermined position or path) and a FOLLOW algorithm (the slave mimics the master's behaviors by a lateral offset.

Yang and Noguchi [64] created a human detection system that used omni-directional stereo vision (OSV) in a 3D panorama image to increase the safety of robot tractor operations. They used two multi-lens-based high-definition omni-directional cameras and two RTK-GPS systems. The position of an obstacle (a human, height 1.7 m) was detected using the RTK-GPS installed on the human's helmet, and this was compared with the results obtained with the OSV system. The RMSE of the distance error and maximum error between the RTK-GPS data and the OSV data were 0.49 and 1 m, respectively. The OSV system also successfully detected the direction of obstacle movement.

Those results demonstrated that the system designed by Yang and Noguchi could detect a mobile or fixed obstacle in the daytime. The system's accuracy was better than those of laser scanners, ultrasonic, and infrared sensors. Using an OSV method could counteract the drawbacks of the sensors mentioned above because of the inconsistency of appearance information, such as the color of the obstacle. Jongmin [259] later improved the performance of a robot tractor and a combine harvester for working during the nighttime.

In 1997, the VeBots laboratory collaborated with the University of Illinois to learn the latter's agricultural robot technology. Zhang and Reid [260] designed a robot tractor [70] designed a robot tractor (AAV-4) that uses DGPS, a FOG sensor, and a camera with a positioning error of <150 mm and a steering error of <0.4° at the traveling speed of 3.6 m/s. They installed navigation sensors, steering sensors, and a control unit on a modified Case-IH 7220 tractor (Case-IH Corp.), as shown in Figure 10.1 (AAV-4). Noguchi and Reid [261] later developed an intelligent vision system for a robot tractor to classify crops. They used fuzzy logic (FL) for crop classification, a genetic algorithm (GA) for the FL optimization of a vision system, an artificial neural network (Aneural network) to estimate the crop height, and GIS for map creation. The system was expected to have a positioning accuracy of 200 mm. The developed system was able to classify the estimated weight and height of soybean crops accurately.

Image processing to detect orientation in which the positioning error was improved to 44.7 mm was reported by Pinto and Reid [262]. The heading error of this guidance algorithm was 1.26°, which is greater than the error afforded by the sensor fusion method. The sensor fusion algorithm of the robot tractor (AAV-4) was improved and recorded by Monte and Noboru [263] as a patent entitled "Sensor-fusion navigator for automated guidance of off-road vehicles" in 2002.

Hokkaido Island is the northernmost island in Japan, with a long winter and an average winter snowfall of 1 m. Only five months of the year can be used to cultivate rice, and the muddy rice fields decrease wheel tractors' efficiency. Takai and Barawid [264] designed a crawler robot tractor (AAV-5) by applying RTK-GPS (Legacy-E, Topcon) with a virtual reference station (VRS) and an IMU (#JCS-7401A, JAEI) on a modified crawler-type tractor (Model CT801, Yanmar, Osaka, Japan), as shown in Figure 10.1 (AAV-5). The least squares method (LSM) was used for the sensor fusion; the GIS provided the navigation map, and the control of each crawler motion was provided by a digital-to-analog converter of the ECU [265]. At the speed of 0.6 m/s, the robot tractor's heading error was 0.22°, and the positioning error was 14 mm, indicating that the maneuvering of this full-crawler robot tractor was more accurate than that of wheel-type robot tractors. It could be because of the specifications of crawlers (which can provide stable motion), or it could be due to the low-speed operation. The positioning error and heading error of the system were as follows: at the speed of 0.6 m/s, 10 mm and 0.24°; at 0.9 m/s, 18 mm and 0.47°; at 1.3 m/s, 29 mm and 0.76°; at 1.7 m/s, 22.3 mm and 0.63°; and at 2 m/s, 18 mm and 0.54° [266].

To increase the safety of the robotic maneuver, three types of safety sensors, i.e., a 2D laser scanner (#LMS291, SICK), a tap switch (#T5-16, Tapeswitch Japan, Matsudo, Japan), and two proximity switch sensors (#ED-130, Keyence, Osaka, Japan) were later used.

In the subsequent research by Barawid and Farrokhi Teimourlou [246] in their efforts to develop an automatic guidance system for orchards in 2008, Barawid and Noguchi [267] extended their studies by developing a small, low-cost robotic vehicle (AAV-6) that uses an electric power source instead of fuel. They used an electric utility vehicle (E-Gator, John Deere Co., Moline, IL) because of its easy maintenance and energy efficiency (Figure 10.1). The E-Gator was modified to have controllable steering, movement, and speed by applying an ECU connected to a personal computer using CAN-bus. A RTK-GPS sensor (#MS750, Trimble) and an IMU (#VN-100, JAEI) were used as positioning and attitude sensors, respectively. A proper motor as an electric engine was selected based on the vehicle's maneuver in on-road and off-road conditions, and then the vehicle's steering was calibrated.

The AAV-6 was evaluated in straight line tests, and the results showed that at the speeds of 0.86, 1.52, and 1.61 m/s, there was a lateral error of 80, 180, and 410 mm and heading error of 1.23°, 3.17°, and 7.67°, respectively. The performance of this robot was compared with that of the robot tractor AAV-3 (MD77) described above, which used a wheel-type tractor. The results indicated that the electric vehicle AAV-6 was more accurate than the fuel robot tractor AAV-3 regarding positioning and attitude. However, charging the electric vehicle's batteries can limit extended applications.

Barawid and Noguchi [268] attached a camera to an AAV-6 to develop a follower vehicle that uses target recognition. The system was observed to follow a mobile target with a heading error of 4.6° and a lateral error of 1 m at the speed of 1.1 m/s [268]. Yin and Noguchi [269] also developed an obstacle detection system using a 3D camera (PMD CamCube 2.0) which had a lateral error of 56 mm and 71 mm in static and dynamic situations, respectively. The maximum positioning accuracy of 100 and 184 mm with an RMSE of 56 and 71 mm was achieved in static and dynamic motion, respectively [270,271].

Ospina and Noguchi [272] used an AAV-6 robot vehicle to estimate dynamic parameters and improve the efficiency of autonomous navigation to reduce the tire sideslip effect. Using bicycle geometrical and dynamic models, the main objective was to model the relationship between the platform's sideslip angle, the yaw rate (as outputs), and the vehicle velocity and steering angle (as inputs). They used potentiometers (#CPP-60, Midori Precisions, Tokyo) on one of the steering wheels to provide the steering angle, plus a microcontroller and a computer for communication and data processing.

The vehicle was tested on a flat concrete surface. The comparison of experimental results and the modeled data confirmed that the developed model could be used for robotic vehicles with an estimate of dynamic parameters; however, some modification was needed because of the non-linearity of the agriculture field conditions. Ospina and Noguchi [273] later evaluated their developed model in different soil conditions, including a flat experimental field, dry soil conditions, and wet soil conditions to find a relation between the tire's lateral forces and slip angles in terms of soil moisture content and cone index. This study was done to improve the automatic steering controller systems of AAVs. The results showed that the model could give a more accurate description of the relation between the tire's lateral forces and slip angles and could be applied to agricultural field conditions with slight modifications.

Yang and Noguchi [274] developed a CAN-bus-based robot tractor (AAV-7) (Figure 10.1). They applied an RTK-GPS sensor (#AGI3, Topcon), an IMU, OSV, and two laser scanners (front and rear, and a 2D laser scanner with 100-mm accuracy [275]) as safety sensors, a personal computer, and two CAN-buses (CAN-1: internal communication between the tractor's ECU and the sensors and actuators, CAN-2: communication between the PC and the robot tractor) on a wheel-type tractor (#EG83, Yanmar).

Three experiments were conducted in a rectangular agricultural field with eight paths with 5-m intervals and a turning algorithm developed by Kise and Noguchi [255]. The experiments were performed at (1) an experimental farm, without implements for an accuracy evaluation (at 1 m/s speed); (2) at a soybean field with a spraying implement for an evaluation of non-straight path navigation (the speed was not reported); and (3) at a standard 2-ha agricultural field for almost 5 hours during the nighttime for a test of the stability of the system for long-term work (speed: 0.6 m/s). In the first, second, and third experiments, the heading and lateral errors were 0.57°, 0.77°, 0.75°, and 50, 20, and 40 mm, respectively. The heading error in the second test was increased because of the weight of the spraying implement. Compared to the previously developed robot tractors (AAV-6 and before), this robot tractor's navigation parameters were improved (AAV-7). The third experiment was also the first reported experiment conducted without daylight.

The efficiency of farmwork could increase if one driver could control two tractors and a robot tractor could follow another human-driven tractor [276,277] developed an algorithm for the AAV-7 robot tractor to follow an AAV-3 tractor operated manually. The robot tractor was equipped with wireless video transmission, four cameras, a laser scanner, and a Bluetooth device. The best distance between the two tractors was 150 m. The experiments were done on six paths, with a U-turning method, a speed of 0.9 m/s, and when the robot tractor moved before the human-driven tractor moved. The results showed that this system had a lateral error (straight line and turning) of 30 mm and a heading error of 0.71° [278]. The lateral error of the human-driven tractor was improved from 60 mm to 30 mm because it followed the robot tractor [279–281].

Zhang and Noguchi [282] developed a robot combined harvester (AAV-8) for different applications (Figure 10.1). They used a commercial combine harvester (Model AG1100, Yanmar) equipped with RTK-GPS+IMU (#AGI3, Topcon). A disadvantage of the AGI3 Topcon RTK-GPS was that it should be used at speeds >3 m/s, which probably decreases the efficiency of autonomous vehicles [283]. After Zhang et al. evaluated the AAV-8 in a straight-line test in the field, the RTK-GPS installation location was modified.

The experimental results demonstrated that the developed robot combine harvester had a heading error of 0.92° and a lateral error of 34.7 mm at the traveling speed of 1.0 m/s. Due to its crawlers, this robot combine harvester was more accurate than the previous robot tractors. Regarding the robot tractor AAV-5, its accuracy was already compared to other robot tractors with similar horsepower values. It is because the crawlers have a more excellent connection surface, which reduces the momentary heading fluctuations.

Zhang et al. later optimized the control parameters by developing a kinematic model for steering and turning functions, and the robot was retested [284]. They reported that the robot's performance was optimized at the speed of 1.2 m/s when the optimized lateral error was 250 mm; however, this lateral error was more significant than those in their earlier report [282]. Rahman and Ishii [285] developed a dynamic model to estimate the heading angle and reduce the harvesting time. They also devised an optimum harvesting area for convex and concave polygon-shaped fields using sensor measurement. The test results indicated that the absolute estimated heading angle was sufficient to apply to crop periphery determinations [286].

Choi and Yin [287] added a laser scanner (#UTM-30LX, Hokuyo USA, Indian Trail, NC) to the AAV-8 combine harvester to provide a navigation system by obtaining more field information (such as the shape of the field). They did not use this sensor for obstacle detection or safety applications. Their findings in static and dynamic (speed: 0.97 m/s) conditions showed that the system had heading errors of 200 and 700 mm and lateral errors of 0.8° and 3°, respectively. This system could distinguish the crop rows and harvested boundary precisely. In the manual operation of the system, the operator can detect the edge of harvester crops and tries to harvest the uncut crops. In the system's autonomous operation, sensor detection was needed to detect the cut/uncut crops and to operate the system to reduce the uncut rows.

In this regard, Teng and Noguchi [288] developed an edge detection system that uses a laser rangefinder that was a laser scanner (#UTM-30LX, Hokuyo) installed on a pan-tilt unit (PTU). The indoor edge detection tests showed an average offset of 217 mm and an RMSE of 6.7 mm. The system had a lateral error of 42.7 and 101.5 mm in static and dynamic experiments in a wheat field. The accuracy of the edge detection in the dynamic condition was affected by the vehicle's vibration, the steering error of the navigation (manually operated in that study), and the nonlinearity of the wheat edges. Because of the laser finder's low resolution, the authors recommended using image processing or sensor fusion in future studies. Fan [289] subsequently developed a real-time quality sensor to predict the protein and moisture of wheat using spectroscopy reflectance; however, this research was not directly related to autonomous navigation [290].

Monitoring paddy field growth and spraying have been another common concern in Hokkaido [232]. Liu and Noguchi [291] developed an unmanned surface vehicle (USV) (AAV-9) for paddy fields. They used a radio-controlled air propeller agricultural vessel (RB-26, Hokuto Yanmar Co., Ebetsu, Japan) as a platform, a personal computer, a GNSS/GPS compass (V100: DGPS + gyroscope, Hemisphere GPS, Calgary, Canada), an ECU (Arduino UNO), servo motors for rudder control, and a magnetic sensor (#GV-101, Futaba, Mobara, Japan) to measure the rotary speed of the engine, and a wireless router for data communication (Figure 10.1).

They evaluated the performance of the USV in three different conditions: (1) line-follow navigation to minimize lateral and heading errors, (2) headland turning to evaluate the turning functions, and (3) map-based navigation. The developed robot USV showed linear and autonomous navigations with lateral errors of 250 and 380 mm, respectively. The USV was also able to navigate ellipses in turning paths. The significant lateral errors were due to a crosswind, and Liu and Noguchi [292] tested an idea to simulate and reduce the effect of a crosswind. Their study's main objective was to determine the USV's maneuverability in paddy fields. The experimental results in a test of a zig-zag maneuver at 1.2 m/s were compared with those of a computer simulation and related parameters. The simulation results indicated that the crosswind force could reach 2.85 N at 3 m/s under 0.1% turbulence. Liu et al. recommended an optimized control algorithm that considers crosswind force be used in future studies [293].

Rice cultivation in paddy fields presents challenges, including maneuverability and labor concerns. Okamoto and his team have been working on the autonomous navigation of a specifically designed rice transplanter by Kubota Co. (Japan) [294–296]. As shown in Figure 10.1 (AAV-10), this specifically designed vehicle has metallic wheels and unique dynamic characteristics that make it easier to control. Okamoto et al. developed a low-cost navigation system that uses an RTK-GNSS sensor (#SPS 855, Trimble) and a CAN-Bus connection. This rice transplanter can maneuver in a paddy field with maximum and minimum lateral errors of 33.6 and 6.7 cm, respectively. Okamoto et al. later equipped the vehicle with a camera (4 mm, Tamron, Saitama, Japan) to recognize pre-provided paths. They observed that they could control the new system by an average lateral error of 10.34 mm. This study is in progress.

As mentioned, autonomous navigation includes the turning phase at the end of each straight maneuver (headland). Several algorithms have been developed to address this: a U-turn, a keyhole turn [266], and a fish-tail turn [248] algorithm. Wang and Noguchi [297] designed a robot tractor (AAV-11) with optimized maneuverability in straight lines and turning at the headland (Figure 10.1). Their main objectives were the optimization of the trajectory, headland distance, and economy indexes. A wheel-type tractor (Model EG105, Yanmar) was equipped with a personal computer, control unit, RTK-GPS sensor (SPS855, Trimble), and IMU (#VN100, VectorNav Technology, Dallas, TX) [112]. Their vehicle's turning function was named "circle back (CB)," which includes two circular motions and one linear motion. Their experiments indicated that the robot tractor could maneuver with lateral and heading errors of 60 mm and 1.2°, respectively, when the platform speed is 1.4 m/s. This turning algorithm provided efficient and optimized maneuvering.

In a continuation of their research into follower robots such as a master-slave system (GoTo and FOLLOW systems) [258] and follower robot tractor [277–279], Zhang and Noguchi [298] developed a leader-follower system that is a collaboration of two independent robot tractors in the field, to reduce the total working time. They equipped a half-crawler tractor (Model EG453, Yanmar) with RTK-GPS (SPS855, Trimble), IMU (VN100, VectorNav), PDA (personal digital assistant), personal computer, and Bluetooth as the robot tractor (AAV-12, Figure 10.1). This system has two major parts: the robot's navigator (all control functions) and the server/client.

The system was first simulated on six paths at the average speed of 0.83 m/s when the distance between the robots was 12 m. The minimum distance between the safety zones was 1.56 m. The total operation time was 22.5 min. When the simulation results satisfied the safe distance criterion, and the system was confirmed to work without collisions, Zhang and Noguchi tested the system in a field using the same parameters as the simulation. The results indicated that the minimum distance between the safety zones was 0.8 m, and the total operation time was 24.5 min. These values were obtained because the software did not include all of the parameters of an actual field and due to IMU errors. The AAV-12 system had 84.3% efficiency compared to a single robot tractor. The lateral errors of the first and second robot tractors were 70 and 50 mm, although the reported lateral error of the system was almost zero.

Zhang and Noguchi [299] then improved the algorithm so that the minimum distance between the safety zones could be decreased to 0.59 m; the total operation time decreased to 20.4 min, and the efficiency increased to 95.1%. As was the case when the VeBots laboratory designed robot tractors with reasonable heading errors, this parameter was not reported in follower robots.

The spark of the multi-robot operation idea was provided by Noguchi and Barawid [300] for rice, wheat, [109] and soybean in Japan, planning a five-year project beginning in 2010. They introduced developed autonomous vehicles (AAV-2, AAV-5, AAV-6, AAV-8, and AAV-10) as platforms, RTK-GNSS and

IMU as navigation sensors, and laser-scanner and ultrasonic sensors as safety sensors as the components of future multi-robotic systems. Their idea was the communication among a robot management system (navigation sensors, sensor fusion, Kalman filter), a real-time monitoring system (management and monitoring), a navigation system, and a safety system using a wireless local-area network (LAN).

After several attempts, Zhang and Noguchi [301] developed the first multi-robot tractor system for agricultural field work. They aimed to increase the efficiency and speed of farm work to decrease the risk of harvesting and other natural limitations such as rain and temperature variation. Their multi-robot system can decrease the economic indexes and labor costs that provide an infrastructure for high-efficiency agriculture. Using a U-shaped turning method, they used the robot tractor (AAV-12) as the leading platform in three patterns: an I-pattern, a V-pattern, and a W-pattern. They controlled the lateral and longitudinal distances between the robots and defined a circular and rectangular safety zone. They defined and evaluated two types of efficiency: the efficiency of the multi-robot working time compared to that of a single robot and the efficiency of multi-robots working time compared to the total consumed time.

As shown in Table 10.2, the I, V, and W patterns were simulated using three robot tractors (18 paths), five robot tractors (30 paths), and seven robot tractors (42 paths). The observed efficiency of these combinations was 268.4%, 365.6%, and 437.8%, respectively, compared to a single robot tractor. The field experiments using three (V-pattern) and four (W-pattern) robot tractors showed efficiency values of 247.5% and 352.9%, respectively, compared to a single robot tractor.

The same study's comparison of working times indicated that the three-robot tractor and four-robot tractor combinations have efficiency and lateral error values of 80% and 82.1% and 50 and 42.5 mm, respectively. Based on the results of safety experiments, the developed system was confirmed to be safe for field performance. The final efficiency evaluation showed that the I-pattern, V-pattern, and W-pattern have 83.2%–89.8%, 67.3%–78.9%, and 59.4%–65.8% ranges of efficiency, respectively, which are strongly dependent on the field size [302].

Heavyweight crops such as pumpkin, watermelon, melon, and cabbage have recently attained remarkable market value in Japan (especially Hokkaido). Still, the number of farmers in Japan continues to decrease because of the strenuous workloads and labor shortage. In a developed country, farming is considered an overwhelming job compared to office work in Japan. The current harvesters are not precise or careful enough, which can damage crops and decline marketing value. Robot technology is a potential solution to these issues, but most of the agricultural robots that are currently available are designed for small-sized and lightweight crops [232].

Vehicle robotics laboratories can design smart robot tractors for natural farms, including farms with heavy crops. Such autonomous vehicles are equipped with different sensors (navigation, safety, and attitude sensors) and different platforms (tractors, combine harvesters, rice transplanters, airboats, and electric utility vehicles). As an intelligent system, a proportionate actuator is necessary as the

TABLE 10.2
The Results of Navigation Evaluation

Project No.	Environment	Speed (m/s)	Heading Error (degree)	Lateral Error (mm)			Navigation Sensor		
				RMSE	Max	Avg	PS	AS	SS
AAV-1	Field	0.5	1	230	510	400	1	1	–
AAV-2	Field	0.5	NA	NA	NA	NA	2	1	9
	Field	0.5	NA	NA	NA	NA	3	1	–
	Field	0.5	NA	NA	NA	NA	3	1	–
	Field	0.5	NA	NA	NA	50	4	1	–
	Field	0.5	NA	NA	NA	NA	4	–	–
	Field	0.6	0.004	38	100	100	5	1,4	–
			0.057			170			
	Orchard	0.5	3.11	60	150	NA	5	4	1
		0.55	3.02	60					
		0.64	3.48	50					
		0.87	4.39	70					
		1.4	3.15	60					
	Field (Flat)	0.8	0.63	65	NA	NA	RTK	2,3	–
	Field (Sloping)	0.64	0.64	30					
	Field (Bumpy)		0.59	37					

(Continued)

TABLE 10.2 (Continued)
The Results of Navigation Evaluation

Project No.	Environment	Speed (m/s)	Heading Error (degree)	Lateral Error (mm)			Navigation Sensor		
				RMSE	Max	Avg	PS	AS	SS
AAV-3	Field	1.5	1.3	60	150	NA	5	4	10
	Field	2.5	NA	30	70	NA	5	4.5	10
	Orchard	0.36	1.5	110	NA	NA	5	5	1,10
	Field	0.86	1.5	NA	NA	110	6	5	2
		1.23	3.3			140			
		1.43	4.9			190			
AAV-4	Field, GNSS-FOG	3.6	0.4	NA	150	NA	U		U
	Field, vision only				200				
	Field	virtual	1.26	41.4	166	44.7	RTK		–
	Field	0.6	0.22	14	NA	NA	6	5	1
AAV-5	Field	0.6	0.24	10	NA	NA	7	5	–
		0.9	0.47	18					
		1.3	0.76	29					
		1.7	0.63	22.3					
		2	0.54	18					
	Field	0.5	0.2	10	NA	NA	6	5	1,4,5

AAV-6	Field	0.86	1.23	80	NA	NA	5	6	10
		1.52	3.17	180	100				
		1.61	7.67	410	184				
AAV-7	Field	static	NA	56	70	NA	6	7	6
		dynamic		71	30				
	Field	1	0.57	50	100	NA	7	U	10,2,7
	Field, implement	–	0.77	20	57.1		8	6	–
	Filed, night	0.6	0.75	40	53	NA	7	U	3,8
	Field	1.4, straight	0.98	41	69				
		1, turning	–	–	200				
	Field	0.5	0.39	29	40	NA	7	6	–
AAV-8	Field	1	0.92	34.7	100	NA	5	7	7
	Field	Static	0.8	20					
		0.97	3	70					
AAV-9	Paddy field	1.2	NA	310	820	230	U	6	–
AAV-10	Paddy field	NA	NA	NA	NA	NA	UD	UD	UD
AAV-11	Field	1.4, straight	1.2	30	160	20	7	U	3.7
		1, turning		10	120	80			
AAV-12	Field	1.16	NA	70	50	40	8	6	7
AAV-13	Field	UD	UD	UD	UD	UD	UD	UD	UD

complementary unit. This actuator could be applied for different applications. Roshanianfard and Noguchi [303] developed a "harvesting robot for heavyweight crops (HRHC)" system [233]. They sought to develop an automatic harvesting system to improve the Japanese farmers' work in different fields. The HRHC system consists of five central units: (1) an automatic robot tractor (which could be an AAV-7, AAV-11, or AAV-12), (2) a 5-degrees-of-freedom (DOF) robotic arm, (3) a specifically designed end-effector, (4) main controlling units, and (5) a vision system (Figure 10.1, AAV-13). Automatic maneuvers in the field, crop detection by a vision system, field monitoring, and precise field operations are only some of the planned applications for the HRHC system [304].

To develop this system, Roshanianfard and Noguchi [305] first designed the development procedure for a low-cost 5-DOF robotic arm for farm use, considering the limitations when facing complex agricultural conditions with a limited workspace and payload. Different parameters, such as joint torque, payload per weight (PPW), and repeatability, were calculated and simulated dynamically and statically. After the system's development, the control algorithm Kamata and Roshanianfard [306] were created using the Denavit and Hartenberg method in forwarding kinematics and inverse kinematics calculations [307]. The performance evaluation of the system revealed that the workspace volume was 8.024×109 mm^3, the front access was 3.518×106 mm^2, and the harvesting area was 808 mm [308].

Before the end-effector of the system was designed, three varieties of pumpkin (JEJEJ, TC2A, Hokutokou) as a heavyweight crop were characterized [233]. Three experiments were conducted: (1) an evaluation of the general physical properties for investigating the pumpkins' orientation and possible harvesting methodologies, (2) a compression strength test for measuring the yield force, and (3) a bending-shear test for measuring the yield force to cut the pumpkin stems. The results showed that the average pumpkin's lift weight was 26% more than the pumpkin's pure weight. A unique applied technique simplified the harvesting procedure [309].

Roshanianfard and Noguchi [310] then designed an end-effector with a unique harvesting methodology based on the extracted properties of pumpkins. The components were designed and dynamically simulated, and after several modifications, the final components were manufactured and assembled. The end-effector had five fingers designed to grasp and harvest heavy-weight crops with a diameter of 170–500 mm by a sustainable force distribution. The trial results demonstrated that the systems' fingers have enough capability under the maximum payload of the HRHC system (i.e., 25 kg), and the new end-effector could harvest the varieties of pumpkin mentioned above because the range of the system's radius, volume, and mass values covered the extracted physical parameters of pumpkins.

After the control unit of the HRHC system was designed (based on a programmable logic controller system consisting of a personal computer, a position board, amplifiers, servo motors, a switching unit, and emergency switches), the performance of the HRHC system was evaluated. The results showed that the HRHC system has a 92% harvesting success rate, 0% damage rate, and an average cycle time of 42.4 s. The final parameters of the workspace were as follows: workspace volume, 5.662×109 mm^3; harvesting surface, 2.86×106 mm^2; and harvesting length, 800 mm. The average accuracy values were as follows: x-direction, 10.91 mm; y-direction, 9.52 mm; and repeatability, 12.74 mm. The control resolution in the x-, y-, and z-directions were 1 ± 0.075, 1 ± 0.05, and 1 ± 0.025, respectively [311].

The harvesting success rate of the HRHC system (100%) was higher than the overall average harvesting success rate of previous studies, 66% (40%–86%) between 1984 and 2012 [312]. The calculation, simulation, development, modification, and characterization results show that the HRHC system can be used for heavyweight crop robotic harvesting applications. Although this system was designed for specific crops (i.e., pumpkin, watermelon, and cabbage), it could be used for other applications such as precise seeding, fertilizing, watering, and weeding.

Table 10.2 presents more detailed information about the applied sensors for each project, the experimental conditions, and the results.

One of the main components of an autonomous vehicle is the mobile platform. For the category of autonomous agriculture vehicles, several platforms can be used. The VeBots laboratory used mostly tractors (in nine projects and 60 papers). They also used a combine harvester (one project, eight papers), a utility vehicle (one project, eight papers), an airboat (one project, four papers), and a rice transplanter (one project, one paper). In general, a tractor was the platform in 70% of the projects (74% of the papers), and the other types of platforms were used in only 30% of the projects and described in 26% of the papers. These data are not surprising because the tractor is a multi-application vehicle that can carry different types of implements and equipment and perform several types of farm work; a tractor can thus be used for various robotic applications.

Among the projects mentioned above, in seven projects (52 papers, 64.2%), a wheel-type transporter system was used; in three projects (14 papers, 17.3%), a half-crawler type was used; and in two projects (11 papers, 13.6%) a crawler transporter was used. Most of the robotic projects thus used a wheel-type tractor, the world's most commonly used agriculture machine.

Table 10.3 summarizes the development procedure of operation functions of AAVs, such as the steering control, forward and backward motion, brake, shift, rotary speed, hitch, and PTO. The advances in steering began with AAV-1, the steering of which controlled using a stepper motor was connected to the steering shaft. An electro-hydraulic valve was used later in AAV-2, –3, –4, –6, –7, –11, and –12. AAV-5 and AAV-8 each had a hydrostatic transmission (HST) system with which the steering could be controlled from a swash plate. The ECU did not direct the steering control of the AAVs until AAV-13. It was because of Japan's strict governmental policy regarding vehicle traffic and autonomous navigation of vehicles; this policy did not allow manufacturers to sell a vehicle with controllable steering. After 2016, the policy was revised, and companies were allowed to offer controllable-steering vehicles. Before that, when an AAV was being designed, some modification of the steering control was required.

TABLE 10.3

Operation Functions of Developed Autonomous Vehicles and Their Development Process

Project No.	Steering	Forward & Backward	Brake	Shift	Rotary Speed	Hitch	PTO
AAV-1	•	×	•	NA	•	NA	NA
AAV-2	•	•	•	×	•	•	×
AAV-3	•	•	•	•	•	•	•
AAV-4	•	•	•	•	•	•	•
AAV-5	•	•	NA	NA	•	•	•
AAV-6	•	•	NA	NA	NA	NA	NA
AAV-7	•	○	○	○	○	○	○
AAV-8	○	○	○	○	○	NA	NA
AAV-9	•	•	•	NA	•	NA	NA
AAV-10	○	○	○	NA	○	NA	NA
AAV-11	•	○	○	○	○	○	○
AAV-12	•	○	○	○	○	○	○
AAV-13	○	○	○	○	○	○	○

○: It can be controlled without modification.
•: It can be controlled by modification.
×: Manual
NA: Not available

The other operation functions (i.e., forward and backward motion, brake, shift, rotary speed, hitch, and PTO control) were performed manually using the related handle. A handle connected to a switch could control the actuator. Until the AAV-6, an extra ECU was required to control these functions. This ECU was connected between a switch and actuators and could control it using the signals from a personal computer. It was why these functions needed further modification. From the AAV-7 onward, manufacturers have included a commercialized ECU on the platforms, making controllable functions possible. After that (except for the AAV-9, which has a different topology, control system, and platform), the operation functions of all AAVs have been controllable without modification.

10.3   THE DEVELOPMENT OF A COMPONENTS

Figure 10.2 illustrates the development procedure of some components including an internal communication system and sensors (positioning sensors, attitude sensor, and safety sensor).

Initially, the internal communication system was directly controlled because no ECU was available on tractors. The research groups were obliged to use different interface cards to receive sensors' data and send commands to the actuators. Some

Component	Progress
Internal communication system	Direct control ▶ RS232c ▶ USB ▶ CAN-Bus
Positioning sensor	Image sensor ▶ LNAV & XNAV ▶ GPS & SNAV ▶ DGPS ▶ RTK-GPS
Attitude sensor	GDS ▶ FOG ▶ IMU
Safety sensor	Switch ▶ Laser scanner ▶ Laser scanner + Camera

FIGURE 10.2   The development procedure of different components.

prototype ECUs were eventually developed and installed on tractors. These ECUs were not commercialized, but researchers and manufacturers investigated several systems to obtain an ideal control unit. At that time, the RS-232c (and a USB in some cases) was used for communication between a personal computer and an ECU. In some systems, a USB was also used. After the ISO-Bus and the CAN-Bus communication systems were developed, the communication of several ECUs became more accessible and faster. Tractors, like cars, can be equipped with CAN-Bus, and this communication method is now constantly used.

When the AAV-1 was developed, GNSS was not yet available, and its positioning sensor had been designed only for military applications. Image sensors were then created to guide a tractor. As the accuracy of this positioning sensor was not good enough, researchers decided to use some available positioning sensors, including LNAV, XNAV, SNAV, and DGPS sensors. The SNAV sensor was selected as the optimal sensor because the navigation topology of SNAV is based on RTK-GPS. However, RTK-GPS and DGPS were still in the early development stages, and their accuracy was not very high. RTK-GPS was also too expensive to use for agricultural applications.

When Japan faced several challenges, such as a labor shortage, the high price of autonomous vehicles became more acceptable, and several laboratories expressed their interest in developing robotic vehicles. During this period, RTK-GPS became the standard position sensor in autonomous vehicles. The current RTK-GPS, with 2 cm accuracy, is an accurate but expensive positioning sensor. The development of low-cost RTK-GPS or a new economic navigation method is necessary.

Overall, attitude sensors can be divided into GDS, FOG, and IMU. GDS was the most commonly used sensor before developing FOG and IMU. GDS and TMS sensors, which are affected by the surrounding magnetic field, were unavoidably inaccurate, resulting in significant random and bias errors. FOG sensors then became more popular, and three-axis FOG sensors were soon used to measure the 3D orientation of vehicles. These sensors were applied in airplane navigation systems before being used in agriculture. Because of the high price of three-axis FOG sensors, micro-electro-mechanical system (MEMS) IMUs were evaluated, and this type of attitude sensor is now frequently used. Thus, attitude sensors will continue to be used in future designs of autonomous agriculture vehicles.

Safety sensors protect AAVs from obstacles in agricultural environments. Contact sensors (e.g., bumper switches) were the first safety sensors. This type of safety sensor has high reliability, but the significant delay and long response time could cause severe damage. A camera was added to the next generation of AAVs to increase the safety index. Different methodologies based on the cameras were applied to analyze the agricultural environment, such as stereo vision. A camera can collect valuable and diverse information about the environment. Still, this type of sensor has two significant disadvantages: (1) more information can reduce safety, and (2) environmental factors such as lights and shadows can give rise to errors. Researchers then started combining a camera with a laser scanner to increase safety.

The camera/laser scanner combination is now the primary choice as a safety sensor, but the use of only a laser scanner has been examined. A laser scanner can cover a longer distance than a camera, and, in most cases, a laser scanner is more accurate. The development of safety sensors began with switches, moved on to cameras, and eventually combined a laser scanner and camera as the most effective sensor.

Agricultural environments have been the primary substrates examined in the development of AAVs for farming, which has provided many challenges in designing robotic systems. This type of substrate is unique because many complex and varying factors must be considered in an agricultural environment. Examples include light reflection, which can affect image sensors; dusty environments, which can affect a mechanical platform or its components; and humidity, rain, wind, non-flat surfaces, isolation, and vibration. These factors increase the design process's complexity, so researchers select a specific environment when designing the required system.

As a conclusion, the heading error in (AAV-2, PS-5, 0.55), (AAV-2, PS-5, 0.6), (AAV-2, PS-5, 0.64), (AAV-2, PS-9, 0.8, Flat), and (AAV-2, PS-9, 0.8, Slope) were $3.11°$, $3.02°$, $3.48°$, $4.39°$, and $3.15°$, respectively. These errors were relatively high and were not suitable for autonomous applications. In standard experimental conditions, the heading error has always been $<2°$. From AAV-3 onward, three-axis FOGs began to be used, and this type of IMU changed over time to the MEMS type.

Some experiments indicated that high speed, a rapid turn, and using only a 2D laser scanner as the positioning sensor could cause significant heading errors. Generally, the average heading errors before (using GDS) and after (using IMU) the threshold were $2°$ and $1.6°$. Using an IMU thus positively affects the heading error in navigation.

The next performance indicator of AAVs is the lateral error, which is directly related to the positioning sensor and the platform, the transporter system, the tire slip angle, and the environmental conditions. Before RTK-GPS, different positioning sensors (e.g., image sensor, LNAV, SNAV, XNAV, and DGPS) were used, and the results were not stable over all different conditions. The lateral error of the AAV-1 was 23 cm—a significant error due to the image sensor and stations. In the first AAV-2 projects, positioning sensors such as LNAV, SNAV, and XNAV were used, with poor results. When the positioning sensor was changed to RTK-GPS, the average lateral error decreased to 6.78 cm. The RTK-GPS, Topcon, and Legacy-E positioning sensors had an average lateral error of 2.5 cm when attached to the AAV-5 and AAV-7. The RTK-GPS, Topcon, and AGI-3 positioning sensors had an average lateral error of 2.61 cm when attached to AAV-5, −7, −8, and −11.

The RTK-GPS, Trimble, and SPS855 sensors had an average lateral error of 6 cm when attached to the AAV-7 and AAV-12. The RTK-GPS, Trimble, and MS750 sensors had an average lateral error of 9.83 cm when attached to the AAV-2, −3, −6, and −8. The DGPS, Hemisphere, and V100 sensors had an average lateral error of 11 cm when attached to the AAV-2 and the AAV-9 (31 cm). Although it has not been possible to determine the lateral error by considering only the positioning sensors, they play a central role in attaining high position accuracy.

Appendix 1: Abbreviations and Nomenclatures

Abbreviation	Explanation
AAV	Autonomous Agricultural Vehicle
AC	Alternating Current
ACS	American Chemical Society
AHRS	Attitude and Heading Reference System
AI	Artificial Intelligence
AIST	Advanced Industrial Science and Technology
ANN	Artificial Neural Network
API	Application Programming
AWD	All-Wheel Drive
BEV	Battery Electric Vehicle
CCD	Charge Coupled Device
CPGPS	Code Phase Global Positioning System
DC	Direct Currency
DGPS	Differential Global Positioning System
DOD	Department of Defense
DVI	Digital Visual Interface
ECU	Engine Control Unit
EGNOS	European Geostationary Navigation Overlay Service
EMS	European Mathematical Society
FAO	Food and Agriculture Organization
FCEV	Fuel Cell Electric Vehicle
FOG	Fiber Optic Gyroscopes
FWD	Front Wheel Drive
GDP	Gross Domestic Product
GDS	Geomagnetic Direction Sensors
GIS	Geographic Information System
GNSS	Global Navigation Satellite System
GPIO	General Purpose Input/Output
GPS	Global Positioning System
GSM	Global System for Mobile
GUI	Graphical User Interface
HEV	Hybrid Electric Vehicle
IAM	International Association of Moers
ICA	Imperialist Competitive Algorithm

ICE	Internal Combustion Engine
ICT	Information and Communications Technology
IDE	Integrated Development Environment
IMU	Inertial Measurement Units
IoT	Internet of Things
ISM	Institute For Supply Management
JVM	Java Virtual Machine
LAN	Local Area Network
LNAV	Lateral Navigation
LSM	Least Squares Method
LTE	Long Term Evolution
MI	Magneto Impedance
MPC	Model Predictive Control
MSAS	Multi-Purpose Satellite Amplification System
MSB	Mediterranean School of Business
OEM	Original Equipment Manufacturer
OSV	Omnidirectional Stereo-Vision
PA-LAB	Precision Agriculture Laboratory
PCB	Printed Circuit Board
PDA	Personal Digital Assistant
PEV	Plug-in Electric Vehicle
PID	Proportional Integral Derivative
PS	Play Station
PSO	Particle Swarm Optimization
PTC	Positive Train Control
PTU	Pan Tilt Unit
RGB	Red Green Blue
RLG	Reverse Logistic Group
ROS	Robot Operating System
RTK	Real-Time Kinematic
RWD	Rear Wheel Drive
SBAS	Satellite-Based Augmentation System
SNAV	Smooth Navigation
TMS	Transcranial Magnetic Stimulation
TRS	Teachers Retirement System
TSP	Thrift Savings Plan
TTS	Text to Speech
UAV	Unmanned Aerial Vehicle
UGV	Unmanned Ground Vehicle
UHF	Ultra-High Frequency
UNDP	United Nations Development Program
UPS	Uninterruptible Power Supply
US	United States

USD	Unite States Dollar
USV	Unmanned Surface Vehicle
UTV	Utility Terrain Vehicle
UUV	Unmanned Underwater Vehicle
WAAS	Wide Area Augmentation System
WIMU	Wireless Inertial Movement Unit

References

1. Goense, D., *The economics of autonomous vehicles in agriculture*. In *2005 ASAE Annual Meeting*. 2005. American Society of Agricultural and Biological Engineers.
2. Rondelli, V., B. Franceschetti, and D. Mengoli, A Review of Current and Historical Research Contributions to the Development of Ground Autonomous Vehicles for Agriculture. *Sustainability*, 2022. **14**(15): p. 9221.
3. Roldán, J.J., et al., Robots in agriculture: State of art and practical experiences. 2018: pp. 67–90. In Antonio Neves J. R. Prof., *Service robots*, InTech.
4. Pedersen, S.M., et al., Agricultural robots—System analysis and economic feasibility. *Precision Agriculture*, 2006. **7**(4): pp. 295–308.
5. Boubin, J., et al., *Autonomic computing challenges in fully autonomous precision agriculture*. In *2019 IEEE international conference on autonomic computing (ICAC)*. 2019. IEEE.
6. Dong, X., M.C. Vuran, and S. Irmak, Autonomous precision agriculture through integration of wireless underground sensor networks with center pivot irrigation systems. *Ad Hoc Networks*, 2013. **11**(7): pp. 1975–1987.
7. Yang, J., et al., A moisture-hungry copper complex harvesting air moisture for potable water and autonomous urban agriculture. *Advanced Materials*, 2020. **32**(39): p. 200-293.
8. Shufeng, H., H. Yong, and F. Hui, Recent development in automatic guidance and autonomous vehicle for agriculture: a review. Journal of Zhejiang University (Agriculture and Life Sciences), 2018. **44**(4): pp. 381–391.
9. Mousazadeh, H.J.J.o.T., A technical review on navigation systems of agricultural autonomous off-road vehicles. *Journal of Terramechanics*, 2013. **50**(3): pp. 211–232.
10. Monarca, D., et al., Autonomous Vehicles Management in Agriculture with Bluetooth Low Energy (BLE) and Passive Radio Frequency Identification (RFID) for Obstacle Avoidance. *Sustainability*, 2022. **14**(15): p. 9393.
11. Anonymous, https://www.worldbank.org/en/topic/agriculture/overview#4
12. Sihaloho, R., T. Prabowo, and R. Kusuma, Food crisis in Yemen: The roles of food and agriculture organization (FAO) from 2015 to 2020. *Journal of International Studies*, 2022. **5**(1): pp. 59–74.
13. Anonymous, https://www.euromonitor.com/global-overview-of-the-agriculture-industry/report
14. Anonymous, https://blog.bizvibe.com/blog/largest-agricultural-companies
15. Anonymous, Monthly magazine of Iran's Agricultural Jihad Organization. 2020.
16. Anonymous, Monthly magazine of Iran's Agricultural Jihad Organization. 2021.
17. Anonymous, https://www.agrocares.com/2020/10/30/what-is-the-difference-between-precision-digital-and-smart-farming/
18. Anonymous, https://www.smart-akis.com/wp-content/uploads/2018/03/Folder_Position_Digitalisierung_e_IT.pdf
19. Magnin, C.J.D.M., How big data will revolutionize the global food chain. *Digital McKinsey*, 2016. https://www.mckinsey.com/capabilities/mckinsey-digital/our-insights/how-big-data-will-revolutionize-the-global-food-chain
20. Giesler, S., Digitization in agriculture-from precision farming to farming 4.0. *BioEconomyBW*, 2019. https://www.biooekonomie-bw.de/en/articles/dossiers/digitisation-in-agriculture-from-precision-farming-to-farming-40

21. Dash, G. and D. Chakraborty, Digital transformation of marketing strategies during a pandemic: Evidence from an emerging economy during COVID-19. *Sustainability,* 2021. **13**(12): p. 6735.

22. Ayush, G. and R. Gowda, A study on impact of covid-19 on digital marketing. *Vidyabharati International Interdisciplinary Research Journal,* 2020: pp. 225–228.

23. Tavasoli, B., et al., *Examining the situation and marketing bottlenecks of Iran's agricultural products. The sixth agricultural economics conference,* Mashhad, Iran, 2007.

24. Alidousti, S., Factors influencing the development of information technology and e-commerce in small and medium-sized companies. *Information processing and management research paper,* 2008. **3**(25): p. 61.

25. Anonymous, www.aerofarms.com

26. Anonymous, www.airwood.in

27. Anonymous, www.alescalife.com

28. Anonymous, www.arable.com

29. Anonymous, www.bitwaterfarms.com

30. Anonymous, www.bluerivertechnology.com

31. Anonymous, www.cropx.com

32. Anonymous, www.ceresimaging.net

33. Anonymous, www.clearpathrobotics.com

34. Anonymous, www.smartdroplet.com

35. Anonymous, www.digitalspringnet.com

36. Anonymous, www.eFarmer.mobi

37. Anonymous, www.freshboxfarms.com

38. Anonymous, www.freightfarms.com

39. Anonymous, www.gardenplug.com

40. Anonymous, www.honeycombcorp.com

41. Anonymous, www.harvestcroo.com

42. Anonymous, www.impossiblefoods.com

43. Anonymous, www.indigoag.com

44. Anonymous, www.orbitalinsight.com

45. Anonymous, www.perfectdayfoods.com

46. Anonymous, www.rachio.com

47. Anonymous, www.skycision.com

48. Anonymous, www.Smart-Ag.com

49. Anonymous, www.terravion.com

50. Anonymous, www.tevatronic.net

51. Anonymous, www.agrimap.com

52. Anonymous, www.bovcontrol.com

53. Anonymous, www.Cheruvu.in

54. Anonymous, www.farmlogs.com

55. Anonymous, www.farmnote.jp

56. Anonymous, www.ganaz.com

57. Anonymous, www.prospera.ag

58. Anonymous, www.stellapps.com

59. Anonymous, www.trecker.com

60. Anonymous, www.tambero.com

61. Anonymous, www.aggrigator.com

62. Anonymous, www.bumpercrop.co

63. Anonymous, www.agriconomie.com

64. Anonymous, www.meicai.cn

65. Anonymous, www.missfresh.cn
66. Anonymous, www.machinio.com
67. Anonymous, www.produceRun.com
68. Anonymous, www.farmersbusinessnetwork.com
69. Anonymous, www.growtheplanet.com
70. Anonymous, www.livestockCity.com
71. Anonymous, www.livestockconnect.com.au
72. Duckett, T., et al., 2018. Agricultural robotics: The future of robotic agriculture, UK-RAS White papers. https://uwe-repository.worktribe.com/output/866226
73. Shamshiri, R., et al., Research and development in agricultural robotics: A perspective of digital farming. *International Journal of Agricultural and Biological Engineering*, 2018.
74. Vougioukas, S., Agricultural robotics. *Annual Review of Control, Robotics, and Autonomous Systems*, 2019. **2**: pp. 365–392.
75. Jadhav, V., et al., Impact of Facebook on leisure travel behavior of Singapore residents. *International Journal of Tourism Cities*, 2018. **4**(2): pp. 155–157.
76. Saleem, M., J. Potgieter, and A Khalidmahmood., Automation in agriculture by machine and deep learning techniques: A review of recent developments. *Precision Agriculture*, 2021. **22**(6): pp. 2053–2091.
77. Zahid, A., et al., Technological advancements towards developing a robotic pruner for apple trees: A review. *Computers and Electronics in Agriculture*, 2021. **189**: p. 106383.
78. Lytridis, C., et al., An Overview of Cooperative Robotics in Agriculture. *Agronomy*, 2021. **11**(9): p. 1818.
79. Atefi, A., et al., Robotic Technologies for High-Throughput Plant Phenotyping: Contemporary Reviews and Future Perspectives. *Frontiers in Plant Science*, 2021. **12**: p. 611940.
80. Martin, P., A Future-Focused View of the Regulation of Rural Technology. 2021.Agronomy. **11**(6): p. 1153.
81. Oliveira, L., A. Moreira, and M. Silva, Advances in agriculture robotics: A state-of-the-art review and challenges ahead. *Robotics*, 2021. **10**(2): p. 52.
82. Fountas, S., et al., Agricultural robotics for field operations. *Sensors*, 2020. **20**(9): p. 2672.
83. Ren, G., et al., Agricultural robotics research applicable to poultry production: A review. *Computers and Electronics in Agriculture*, 2020. **169**: p. 105216.
84. Zhao, Y., et al., A review of key techniques of vision-based control for harvesting robot. *Computers and Electronics in Agriculture*, 2016. **127**: pp. 311–323.
85. Raibert, M., H. Brown J., and M. Chepponis, Experiments in balance with a 3D one-legged hopping machine. *The International Journal of Robotics Research*, 1984. **3**(2): pp. 75–92.
86. Ahmadi, M., and M. Buehler, Stable control of a simulated one-legged running robot with hip and leg compliance. *IEEE Transactions on Robotics and Automation*, 1997. **13**(1): pp. 96–104.
87. Hirose, S., and Manipulators, Biologically Inspired Robots: Snake-Like Locomotors and Manipulators, 1993. Shigeo Hirose Oxford University Press
88. Unver, O., et al., *Geckobot: A gecko inspired climbing robot using elastomer adhesives*. In *Proceedings 2006 IEEE International Conference on Robotics and Automation, 2006. ICRA 2006*. 2006. IEEE.
89. Kovac, M., et al., *A miniature 7g jumping robot*. In *2008 IEEE international conference on robotics and automation*. 2008. IEEE.

90. Zaitsev, V., et al., A locust-inspired miniature jumping robot. *Bioinspiration & Biomimetics*, 2015. **10**(6): p. 066012.

91. Hu, H., *Biologically inspired design of autonomous robotic fish at Essex*. In *IEEE SMC UK-RI Chapter Conference, on Advances in Cybernetic Systems*. 2006. Citeseer.

92. Erdman, D.J. and D.J. Rose, *A newton waveform relaxation algorithm for circuit simulation*. In *1989 IEEE International Conference on Computer-Aided Design*. 1989. IEEE Computer Society.

93. Roshanianfard, A., and Noboru Noguchi, Pumpkin harvesting robotic end-effector. *Computers and Electronics in Agriculture*, 2020. **174**: p. 105503.

94. Noguchi, N., K. Ishii, and H. Terao, A study on Intelligent Industrial vehicle by using Neural Network (Modeling Vehicle Movement by Neural Network). In *Proceedings of the 2nd Intelligent Systems Symposium Proceedings*. 1992: Japan. pp. 367–372.

95. List, J. *A new open-source farming robot takes shape*. 2021; Available from: https://hackaday.com/2021/03/05/a-new-open-source-farming-robot-takes-shape/

96. Fue, K., et al., Center-articulated hydrostatic cotton harvesting rover using visual-servoing control and a finite state machine. *Electronics*, 2020. **9**(8): p. 1226.

97. Madokoro, H., et al., Prototype development of small mobile robots for mallard navigation in paddy fields: Toward realizing remote farming. *Robotics*, 2021. **10**(2): p. 63.

98. Yanmar Co. Data sheet for YT488A, YT498A, YT4104A, and YT5113A tractors. 2018.

99. Boston Dynamics. The mobile robot designed for sensing, inspection, and remote operation (Spot). 2019; Available from: https://www.bostondynamics.com/

100. DJI, Co. 2023. https://nauav.ir/pdf/DJI-Agriculture-Case-Study-Manual-2021-negah-aseman.pdf

101. Ouezdou, F.B., S. Alfayad, and B. Almasri, *Comparison of several kinds of feet for humanoid robot*. In *5th IEEE-RAS International Conference on Humanoid Robots, 2005*. 2005. IEEE.

102. Ishii, K., H. Terao, and N. Noguchi, Studies on Self-learning Autonomous Vehicles (Part 3) positioning system for Autonomous Vehicle. *Journal of the Japanese Society of Agricultural Machinery*, 1998. **60**(1): pp. 51–58.

103. Kondo, N., M. Monta, and N. Noguchi, Agricultural Robots: Mechanisms and Practice, 2006. N. Noguchi . Prof. Trans Pacific Press.

104. El-Rabbany, A., Introduction to GPS: the global positioning system. 2002. Artech House.

105. Wanninger, L., Introduction to network RTK. *IAG Working Group*, 2004. **4**(1): pp. 2003–2007.

106. Mannings, R., *Ubiquitous positioning*. 2008. Artech House.

107. Weiffenbach, G., Tropospheric and ionospheric propagation effects on satellite radio-doppler geodesy. *Hilger and Watts*, 1967: p. 339.

108. RietDorf, A., C. Daub, and P. Loef. *Precise positioning in real-time using navigation satellites and telecommunication*. In *Proceedings of The 3rd Workshop on Positioning and Communication (WPNC'06)*. 2006. Citeseer.

109. Yoshihisa, Kawamura, et al., Geomagnetic direction sensor, U. States, Editor. *European Patent Office*, 1995.

110. Neubrex, T. *Fiber Optic Gyroscope development*. 2007; Available from: https://www.neubrex.com/htm/applications/gyro-principle.htm

111. NEDAERO. *Fiber Optic Gyroscope (FOG)*. 2022; Available from: https://nedaero.com/fog/

112. VectorNAV. VN-100, Introduction. 2022; Available from: https://www.
     vectornav.com/products/detail/vn-100

113. Grady, J.O., *System requirements analysis*. 2010. Elsevier.

114. Nayyar, A. and V. Puri, *A review of Arduino board's, Lilypad's & Arduino
     shields*. In *2016 3rd International Conference on Computing for Sustainable
     Global Development (INDIACom)*. 2016. IEEE.

115. Sobota, J., et al., Raspberry Pi and Arduino boards in control education. *IFAC
     Proceedings*, 2013. **46**(17): pp. 7–12.

116. Karthikeyan, P., et al., *Design and Development of Laplacian Pyramid Combined
     with Bilateral Filtering Based Image Denoising*. In *International Conference on
     Soft Computing Systems*. 2018. Springer.

117. Anonymous, https://www.elprocus.com/beaglebone-black-microcontroller/

118. Anonymous, https://www.dummies.com/article/technology/computers/hardware/
     beaglebone/comparing-beaglebone-black-and-raspberry-pi-145670/

119. Anonymous, https://www.robotpark.com/Adafruit-Board

120. Horsey, J., *Adafruit unveils new circuit playground board to learn about elec-
     tronics*. 2016. Geeky Gadgets.

121. Anonymous, https://www.javatpoint.com/what-is-computer

122. Anonymous, https://opentextbc.ca/computerstudies/chapter/types-of-
     computers/#:~:text=Supercomputers,Personal%20computers%20(PCs)%20or
     %20microcomputers

123. Jasch, Christine, Environmental performance evaluation and indicators. *Journal
     of Cleaner Production*, 2000. **8**(1): pp. 79–88.

124. Azzone, G., et al., Defining environmental performance indicators: Anintegrated
     framework. *Business Strategy and the Environment*, 1996. **5**(2): pp. 69–80.

125. Fatima, S. and M. Seshashayee, *Hybrid using*. In *2020 3rd International
     Conference on Intelligent Sustainable Systems (ICISS)*. 2020. IEEE.

126. Tu, S., et al., Passion fruit detection and counting based on multiple scale faster
     R-CNN using RGB-D images. *Precision Agriculture*, 2020. **21**(5): pp. 1072–1091.

127. Huang, Yo-Ping, Tzu-Hao Wang, and Haobijam Basanta, Using fuzzy mask
     R-CNN model to automatically identify tomato ripeness. *IEEE Access*, 2020. **8**:
     pp. 207672–207682.

128. Patel, J., B. Vala, and M. Saiyad, LSTM-RNN Combined Approach for Crop
     Yield Prediction On Climatic Constraints. In *2021 5th International Conference
     on Computing Methodologies and Communication (ICCMC)*. 2021. IEEE.

129. Yang, M.-D., et al., Assessment of grain harvest moisture content using machine
     learning on smartphone images for optimal harvest timing. *Sensors*, 2021. **21**(17):
     p. 5875.

130. Najafi, B., et al., Application of ANNs, ANFIS and RSM to estimating and optimizing
     the parameters that affect the yield and cost of biodiesel production. *Engineering
     Applications of Computational Fluid Mechanics*, 2018. **12**(1): pp. 611–624.

131. Behrang, M., et al., The potential of different artificial neural network (ANN)
     techniques in daily global solar radiation modeling based on meteorological data.
     Solar Energy, 2010. **84**(8): pp. 1468–1480.

132. McCulloch, W.S. and P. Walter, A logical calculus of the ideas immanent in
     nervous activity. *The Bulletin of Mathematical Biophysics*, 1943. **5**(4):
     pp. 115–133.

133. Faizollahzadeh Ardabili, S., et al., Fuzzy logic method for the prediction of cetane
     number using carbon number, double bounds, iodic, and saponification values of
     biodiesel fuels. *Environmental Progress & Sustainable Energy*, 2019. **38**(2):
     pp. 584–599.

134. Yilmaz, I., and A. Yuksek, An example of artificial neural network (ANN) application for indirect estimation of rock parameters. *Rock Mechanics and Rock Engineering*, 2008. **41**(5): p. 781.
135. Agatonovic-Kustrin, S., and R. Beresford, Basic concepts of artificial neural network (ANN) modeling and its application in pharmaceutical research. *Journal of Pharmaceutical and Biomedical Analysis*, 2000. **22**(5): pp. 717–727.
136. Elmolla, E.S., M. Chaudhuri, and M. Eltoukhy, The use of artificial neural network (ANN) for modeling of COD removal from antibiotic aqueous solution by the Fenton process. *Journal of Hazardous Materials*, 2010. **179**(1–3): pp. 127–134.
137. Hara, P., M. Piekutowska, and G. Niedbała, Selection of independent variables for crop yield prediction using artificial neural network models with remote sensing data. *Land*, 2021. **10**(6): p. 609.
138. Wickramasinghe, L., J. Jayasinghe, and U. Rathnayake. *Relationships between climatic factors to the paddy yield in the north-western province of Sri Lanka*. In *2020 International Research Conference on Smart Computing and Systems Engineering (SCSE)*. 2020. IEEE.
139. Savarapu, P.R., et al., *Enhanced computerized classification system of diseased leaves*. In *2020 4th International Conference on Electronics, Communication and Aerospace Technology (ICECA)*. 2020. IEEE.
140. Shad, M., et al., Forecasting of monthly relative humidity in Delhi, India, using SARIMA and ANN models. *Modeling Earth Systems and Environment*, 2022: pp. 1–9.
141. Rashvand, M. and M. S. Firouz, Dielectric technique combined with artificial neural network and support vector regression in moisture content prediction of olive. *Research in Agricultural Engineering*, 2020. **66**(1): pp. 1–7.
142. Akhand, K., et al., *Using remote sensing satellite data and artificial neural network for prediction of potato yield in Bangladesh*. In *Remote Sensing and Modeling of Ecosystems for Sustainability XIII*. 2016. SPIE.
143. Ravichandran, G. and R. Koteeshwari, *Agricultural crop predictor and advisor using ANN for smartphones*. In *2016 International Conference on Emerging Trends in Engineering, Technology and Science (ICETETS)*. 2016. IEEE.
144. Ganti, R. *Monthly monsoon rainfall forecasting using artificial neural networks*. In *AIP Conference Proceedings*. 2014. American Institute of Physics.
145. Safa, M. and S. Samarasinghe, Modelling fuel consumption in wheat production using artificial neural networks. *Energy*, 2013. **49**: pp. 337–343.
146. Fadilah, N., et al., Intelligent color vision system for ripeness classification of oil palm fresh fruit bunch. *Sensors*, 2012. **12**(10): pp. 14179–14195.
147. Kandel, A., *Fuzzy mathematical techniques with applications*. 1986. Addison-Wesley Longman Publishing.
148. Tanaka, K. and K. Tanaka, *An introduction to fuzzy logic for practical applications*. 1997. Springer.
149. Zadeh, L.A., G.J. Klir, and B. Yuan, *Fuzzy sets, fuzzy logic, and fuzzy systems: Selected papers*. Vol. 6. 1996. World Scientific.
150. Wang, L.-X., *Adaptive fuzzy systems and control: design and stability analysis*. 1994. Prentice-Hall.
151. Wang, P.P., D. Ruan, and E.E. Kerre, *Fuzzy logic: A spectrum of theoretical & practical issues*. Vol. 215. 2007. Springer.
152. Pasgianos, G., et al., A nonlinear feedback technique for greenhouse environmental control. *Computers and Electronics in Agriculture*, 2003. **40**(1–3): pp. 153–177.

153. Bai, Y., H. Zhuang, and D. Wang, *Advanced fuzzy logic technologies in industrial applications.* 2006. Springer.
154. Ruffoni, E.P., Capabilities and innovative performance in the Brazilian agricultural machinery industry. *International Journal of Business and Management*, 2022. **24**: pp. 275–293.
155. Runtao, W. , et al., Design and experiment of speed-following variable spray system based on fuzzy control. *Nongye Jixie Xuebao/Transactions of the Chinese Society of Agricultural Machinery*, 2022. **53**(6).
156. Smania, G.S., et al., The relationships between digitalization and ecosystem-related capabilities for service innovation in agricultural machinery manufacturers. 2022. **343**: p. 130982.
157. Sharma, R.P., et al., IoT-enabled IEEE 802.15. 4 WSN monitoring infrastructure-driven fuzzy-logic-based crop pest prediction. *IEEE Internet of Things Journal*, 2021. **9**(4): pp. 3037–3045.
158. Ji, K., et al., Device and method suitable for matching and adjusting reel speed and forward speed of multi-crop harvesting. *Agriculture*, 2022. **12**(2): p. 213.
159. Mansour, T., *PID control: Implementation and turning.* 2011. BoD–Books on Demand.
160. Panda, R.C., *Introduction to PID controllers: Theory, tuning and application to frontier areas.* 2012. BoD–Books on Demand.
161. Zhou, M., et al., Development of a depth control system based on variable-gain single-neuron PID for rotary burying of stubbles. *Agriculture*, 2021. **12**(1): p. 30.
162. Jin, X., et al., Simulation of hydraulic transplanting robot control system based on fuzzy PID controller. *Measurement*, 2020. **164**: p. 108023.
163. Siddique, M.A.A., et al., *Simulation of hydraulic system of the rice transplanter with AMESim software.* In *2018 ASABE Annual International Meeting.* 2018. American Society of Agricultural and Biological Engineers.
164. Zhai, Z., et al., Test of binocular vision-based guidance for tractor based on virtual reality. *Transactions of the Chinese Society of Agricultural Engineering*, 2017. **33**(23): pp. 56–65.
165. Ding, Y., et al., Model predictive control and its application in agriculture: A review. *Computers and Electronics in Agriculture*, 2018. **151**: pp. 104–117.
166. Liu, H., et al., Model predictive control system based on direct yaw moment control for 4WID self-steering agriculture vehicle. *International Journal of Agricultural and Biological Engineering*, 2021. **14**(2): pp. 175–181.
167. Sebastian, R.J. and D. Patino. *Model control predictive of a cold room for an alture application.* In *2019 IEEE 4th Colombian Conference on Automatic Control (CCAC).* 2019. IEEE.
168. Zhang, D., et al., Experimental investigation on model predictive control of radiant floor cooling combined with underfloor ventilation system. *Energy*, 2019. **176**: pp. 23–33.
169. Yang, Z., et al., *On the single-zone modeling for optimal climate control of a real-sized livestock stable system.* In *2009 International Conference on Mechatronics and Automation.* 2009. IEEE.
170. Kružić, S., et al., End-Effector Force and Joint Torque Estimation of a 7-DoF Robotic Manipulator Using Deep Learning. *Electronics*, 2021. **10**(23): p. 2963.
171. Pramod, A.S. and T. Jithinmon, Development of mobile dual PR arm agricultural robot. In *Journal of Physics: Conference Series.* 2019. IOP Publishing.
172. Bogue, Robert, Robots poised to revolutionise agriculture. *Industrial Robot: An International Journal*, 2016. **43**(5): pp. 450–456.

173. Nikolaev, M., I. Nesmianov, and E. Zaharov, *Definition of service area of agricultural loading robot with manipulator of parallel-serial structure*. In *IOP Conference Series: Materials Science and Engineering*. 2020. IOP Publishing.

174. Sakai, S., M. Iida, and M. Umeda. *Heavy material handling manipulator for agricultural robot*. In *Proceedings 2002 IEEE International Conference on Robotics and Automation (Cat. No. 02CH37292)*. 2002. IEEE.

175. Anonymous, http://www.ssl.umd.edu/projects/rangertsx/data/spacerobotics-UNDSPST470.pdf

176. Anonymous, https://www.osha.gov/otm

177. Anonymous, https://robotsdoneright.com/Articles/what-is-an-articulated-robot.html

178. Robots, D., *ABB IRB 6600-225/2.55*. 2022; Available from: https://robotsdoneright.com/ABB/6000-Series/ABB-IRB-6600-225-2.55.html

179. Moscoso, C., E. Sorogastúa, and R. Gardini, Efficient Implementation of a Cartesian Farmbot Robot for Agricultural Applications in the Region La Libertad-Peru. In *2018 IEEE ANDESCON*. 2018. IEEE.

180. Festo, *Three-dimensional gantry YXCR*. 2022.

181. Yamaha. *Clean cartesian robots XY-XC type*. 2022; Available from: https://www.renexrobotics.pl/en/clean-robots/clean-cartesian-robots/

182. Shariatee, M., et al., *Design of an economical SCARA robot for industrial applications*. In *2014 Second RSI/ISM International Conference on Robotics and Mechatronics (ICRoM)*. 2014. IEEE.

183. Yamaha. *Large type SCARA robots YK-XG*. 2022; Available from: https://global.yamaha-motor.com/business/robot/lineup/ykxg/large/

184. FANUC. *SCARA Robot SR-3iA*. 2022; Available from: https://www.fanuc.eu/rs/en/robots/robot-filter-page/scara-series/scara-sr-3ia

185. Brinker, J., et al., Comparative study of serial-parallel delta robots with full orientation capabilities. *IEEE Robotics and Automation Letters*, 2017. **2**(2): pp. 920–926.

186. FANUC. *M-1iA/0.5S*. 2022; Available from: https://www.fanuc.eu/hu/en/robots/robot-filter-page/delta-robots/m1-series/m-1ia-05s

187. ABB. *Parallel Robot IRB 360 FlexPicker*. 2022; Available from: https://www.directindustry.com/prod/abb-robotics/product-30265-169123.html

188. Edan, Y., et al., Near-minimum-time task planning for fruit-picking robots. *IEEE Transactions on Robotics and Automation*, 1991. **7**(1): pp. 48–56.

189. PISHROBOT. *ARMC6MX28*. 2022; Available from: https://www.pishrobot.com/en/product-en/pishrobot-arms/

190. Warrier, A.S. and P. Minz, Robotics in Dairydairy & Food Industryfood industry. In *9th Convention of India Dairy Engineers and IDEA National Seminar*. 2014, National Dairy Research Institute.

191. Vrochidou, E., et al., An overview of end effectors in agricultural robotic harvesting systems. *Agriculture*, 2022. **12**(8): p. 1240.

192. Vu, Q., M. Kuzov, and A. Ronzhin. *Hierarchical classification of robotic grippers applied for agricultural object manipulations*. In *MATEC Web of Conferences*. 2018. EDP Sciences.

193. Longsheng, F., et al., Development and experiment of end-effector for kiwifruit harvesting robot. *Nongye Jixie Xuebao/Transactions of the Chinese Society of Agricultural Machinery*. 2015, **46**(3).

194. Monta, M., N. Kondo, and K. Ting, *End-effectors for tomato harvesting robot*. In *Artificial Intelligence for Biology and Agriculture*. 1998, Springer. pp. 1–25.

195. Zhao, Y., et al., Robust tomato recognition for robotic harvesting using feature images fusion. *Sensors*, 2016. **16**(2): p. 173.

196. Kondo, N., et al., Visual feedback guided robotic cherry tomato harvesting. *Transactions of the ASAE*, 1996. **39**(6): pp. 2331–2338.
197. De-An, Z., et al., Design and control of an apple harvesting robot. *Biosystems Engineering*, 2011. **110**(2): pp. 112–122.
198. Van Henten, E.J., et al., An autonomous robot for harvesting cucumbers in greenhouses. *Autonomous Robots*, 2002. **13**(3): pp. 241–258.
199. Feng, Q., et al., Design and test of robotic harvesting system for cherry tomato. *International Journal of Agricultural and Biological Engineering*, 2018. **11**(1): pp. 96–100.
200. Mu, L., et al., Design and simulation of an integrated end-effector for picking kiwifruit by robot. *Information Processing in Agriculture*, 2020. **7**(1): pp. 58–71.
201. Huang, P., et al., Row end detection and headland turning control for an autonomous banana-picking robot. *Machines*, 2021. **9**(5): p. 103.
202. Vrochidou, E., et al., An Overview of End Effectors in Agricultural Robotic Harvesting Systems. *Agriculture*, 2022. **12**(8): p. 1240.
203. Roshanianfard, A. and N. Noguchi, Pumpkin harvesting robotic end-effector. *Computers and Electronics in Agriculture*, 2020. **174**: p. 105503.
204. Gao, J., et al., Development and evaluation of a pneumatic finger-like end-effector for cherry tomato harvesting robot in greenhouse. *Computers and Electronics in Agriculture*, 2022. **197**: p. 106879.
205. Anonymous, http://myrobotlab.org/
206. Anonymous, www.npco.net *(In persian)*.
207. Anonymous, www.microsoftme.net/what-is-visual-studio *(In persian)*.
208. Anonymous, https://digispark.ir/processing-and-arduino-ide-program/ *(In persian)*.
209. Anonymous, https://www.beytoote.com/computer/tarfand-c/windows-notepad-application.html
210. Anonymous, https://irenx.ir/arduino/arduino-software-for-android/
211. Anonymous, https://faradars.org/courses/fvmec9809-introduction-to-robot-operating-system
212. Newman, W.S., *A systematic approach to learning robot programming with ROS*. 2017. Chapman and Hall/CRC.
213. Anonymous, *"ROS/Introduction – ROS Wiki". ROS.org. Open Robotics. Retrieved 30 July 2021.*
214. Kay, Jackie, Proposal for implementation of real-time systems in ROS 2. *ROS. org*, 2016.
215. Maruyama, Y., S. Kato, and T. Azumi, *Exploring the performance of ROS2. In Proceedings of the 13th International Conference on Embedded Software*. 2016. Association for Computing Machinery.
216. Anonymous, *"rosjava – ROS Wiki". ROS.org. Open Robotics. Retrieved 29 April 2019.*
217. Anonymous, http://wiki.iranros.com/tutorial/intermediate/gazebo/
218. Anonymous, *"About– Ignition Robotics".* www.ignitionrobotics.org. *Retrieved 5 April 2022.*
219. Anonymous, https://irsanat.com/
220. Anonymous, https://king3d.ir/ *(In persian)*.
221. Bryant, S., https://donyad.com/ *(In persian)*.
222. Sariasan, https://sariasan.com/solidworks-tutorial/solidworks-introduction-and-timing/ *(In persian)*.
223. Sariasan, https://sariasan.com/autocad-tutorial/starter/what-is-autocad/ *(In persian)*.
224. Bamdad, https://www.bamdad.co/mag/what-i-catia/. 2019.

225. Anonymous, https://3dmaxclass.com/sketchup.aspx
226. Zarei, S., https://melec.ir/getting-started-with-fritzing-software/
227. https://www.falstad.com/circuit
228. Andish, B., https://behsanandish.com/learning/opencv *(In persian)*.
229. Zerang, A., http://roboticholor.blogfa.com/post/20 (In persian), 2017.
230. Faradars, https://faradars.org/courses/pcb-design-in-eagle-fvee010 *(In persian)*.
231. Roshanianfard, A., et al., A review of autonomous agricultural vehicles (The experience of Hokkaido University). *Journal of Terramechanics*, 2020. **91**: pp. 155–183.
232. Roshanianfard, A., *YouTube Channel named "Agricultural Robots"*. Laboratory of Vehicle Robotics at Hokkaido University – Japan 2020; Available from: https://www.youtube.com/channel/UCRxZvgI9hEGAMsAz8jmWGTQ/playlists?view_as=subscriber
233. Roshanianfard, A., *Development of a harvesting robot for heavy-weight crop*. In *Department of Environment Resources in the Graduate school of Agriculture*. 2018, Hokkaido University. p. 236.
234. Noguchi, N., K. Ishii, and H. Terao, Development of an agricultural mobile robot using a geomagnetic direction sensor and image sensors. *Journal of Agricultural Engineering Research*, 1996. **67**(1): pp. 1–15.
235. Ishii, K., *Study on control method of agricultural autonomous mobile robot*. In *Department of Environment Resources, Graduate school of Agriculture*. 1997, Hokkaido University. p. 146.
236. Noguchi, N. and H. Terao, Path planning of an agricultural mobile robot by neural network and genetic algorithm. *Computers and Electronics in Agriculture*, 1997. **18**(2): pp. 187–204.
237. Ishii, K., H. Terao, and N. Noguchi, Studies on self-learning autonomous vehicles (Part 1), Following control using neuro-controller. *Journal of the Japanese Society of Agricultural Machinery*, 1994. **56**(4): pp. 53–60.
238. Ishii, K., H. Terao, and N. Noguchi, Studies on self-learning autonomous vehicles (Part 2), Verification of neuro-controller by model vehicle. *Journal of the Japanese Society of Agricultural Machinery*, 1995. **57**(6): pp. 61–67.
239. Yukumoto, O., Y. Matsuo, and N. Noguchi, Robotization of agricultural vehicles (Part 1), Component technologies and navigation systems. *Japan Agricultural Research Quarterly*, 2000. **34**(2): pp. 99–105.
240. Yukumoto, O., Y. Matsuo, and N. Noguchi, Robotization of agricultural vehicles (Part 2), Description of the tilling robot. *Japan Agricultural Research Quarterly*, 2000. **34**(2).
241. Anonymous, *Kubota Tractor Corporation. Kubota Co.*, 2018.
242. Mizushima, A., et al., Automatic guidance system composed of geomagnetic direction sensor and fiber optic gyroscope. *IFAC Proceedings Volumes*, 2000. **33**(29): pp. 313–317.
243. Mizushima, A., et al., Development of a low-cost attitude sensor for agricultural vehicles. *Computers and Electronics in Agriculture*, 2011. **76**(2): pp. 198–204.
244. Liu, Y., N. Noguchi, and K. Ishii, Development of a low-cost IMU by using sensor fusion for attitude angle estimation. *IFAC Proceedings Volumes*, 2014. **47**(3): pp. 4435–4440.
245. Tsubota, R., N. Noguchi, and A. Mizushima. *Automatic guidance with a laser scanner for a robot tractor in an orchard*. In *Automation Technology for Off-Road Equipment, Proceedings of the 7–8 October 2004 Conference (Kyoto, Japan) Publication Date 7 October 2004*. 2004. St. Joseph, MI: ASABE.

246. Barawid Jr, O.C., et al., Automatic guidance system in real-time orchard application (Part 1), A novel research on coconut field application using laser scanner. *Journal of the Japanese Society of Agricultural Machinery*, 2008. **70**(6): pp. 76–84.

247. Kise, M., et al., Development of the agricultural autonomous tractor with an RTK-GPS and a fog. *IFAC Proceedings Volumes*, 2001. **34**(19): pp. 99–104.

248. Michio, K., et al., Field mobile robot navigated by RTK-GPS and FOG (Part 2) – Autonomous operation by applying navigation map. *Journal of Agricultural Mechanical Engineering*, 2001. **63**(5): pp. 80–85.

249. Noguchi, N., et al., *Development of robot tractor based on RTK-GPS and gyroscope*. In *2001 ASAE Annual Meeting*. 2001, St. Joseph, MI: ASAE.

250. Noguchi, N., et al., *Field Automation Using Robot Tractor*. In *Automation Technology for Off-Road Equipment, Proceedings of the July 26–27, 2002 Conference (Chicago, Illinois, USA)*, pp. 239–245. 2002. St. Joseph, MI: ASABE.

251. Mizushima, A., et al., *Automatic Navigation of the Agricultural Vehicle by the Geomagnetic Direction Sensor and Gyroscope*. In *Automation Technology for Off-Road Equipment, Proceedings of the July 26–27, 2002 Conference (Chicago, Illinois, USA), pp. 204–211*. 2002. St. Joseph, MI: ASABE.

252. Mizushima, A., N. Noguchi, and K. Ishii, *Development of robot tractor using the low-cost GPS/INS system*. 2005. St. Joseph, MI: ASAE.

253. Noguchi, N., et al., *Turning function for robot tractor based on spline function*. In *2001 ASAE Annual Meeting*. 2001, St. Joseph, MI: ASAE.

254. Kise, M., et al., *The development of the autonomous tractor with steering controller applied by optimal control*. In *Automation Technology for Off-Road Equipment, Proceedings of the July 26–27, 2002 Conference (Chicago, Illinois, USA), pp. 367–373*. 2002. St. Joseph, MI: ASABE.

255. Kise, M., et al., *Enhancement of turning accuracy by path planning for robot tractor*. In *Automation Technology for Off-Road Equipment, Proceedings of the July 26–27, 2002 Conference (Chicago, Illinois, USA), pp. 398–404*. 2002. St. Joseph, MI: ASABE.

256. Yokota, M., et al., *3-D GIS map generation using a robot tractor with a laser scanner*. 2005, St. Joseph, MI: ASAE.

257. Barawid Jr, O.C., K. Ishii, and N. Noguchi, *Calibration method for 2-dimensional laser scanner attached on a robot vehicle*. In *Proceedings of the 17th World Congress, The International Federation of Automatic Control*. 2008. Seoul, Korea.

258. Noguchi, N., et al., Development of a master–slave robot system for farm operations. *Computers and Electronics in Agriculture*, 2004. **44**(1): pp. 1–19.

259. Jongmin, C., *Development of guidance system using local sensors for agricultural vehicles*. In *Department of Environment Resources, Graduate school of Agriculture*. 2014, Hokkaido University.

260. Zhang, Q., J.F. Reid, and N. Noguchi, *Agricultural vehicle navigation using multiple guidance sensors*. In *International Conference on Field and Service Robotics*. 1999.

261. Noguchi, N., et al., Vision Intelligence for Mobile Agro-Robotic System. *Journal of Robotics and Mechatronics*, 1999. **11**(3): pp. 193–199.

262. Pinto, F.A.C., et al., Vehicle guidance parameter determination from crop row images using principal component analysis. *Journal of Agricultural Engineering Research*, 2000. **75**(3): pp. 257–264.

263. Monte, A.D., et al., *Sensor-fusion navigator for automated guidance of off-road vehicles. Journal of Terramechanics*, 2000. **50**: pp. 211–232.
264. Takai, R., et al., Development of crawler-type robot tractor based on GPS and IMU. *IFAC Proceedings Volumes*, 2010. **43**(26): pp. 151–156.
265. Takai, R., L. Yang, and N. Noguchi, Development of a crawler-type robot tractor using RTK-GPS and IMU. *Engineering in Agriculture, Environment and Food*, 2014. **7**(4): pp. 143–147.
266. Takai, R., O. Barawid, and N. Noguchi, Autonomous navigation system of crawler-type robot tractor. *IFAC Proceedings Volumes*, 2011. **44**(1): pp. 14165–14169.
267. Barawid Jr, O.C. and N. Noguchi, Automatic Guidance system in real-time orchard application (Part 2), Development of low-cost and small scale electronic robot vehicle for orchard application. *Journal of the Japanese Society of Agricultural Machinery*, 2010. **72**(3): pp. 243–250.
268. Barawid, O.C. and N. Noguchi, Automatic steering system for electronic robot vehicle. *IFAC Proceedings Volumes*, 2011. **44**(1): pp. 2901–2906.
269. Yin, X., N. Noguchi, and K. Ishi, Development of an obstacle avoidance system for a field robot using a 3D camera. *Engineering in Agriculture, Environment and Food*, 2013. **6**(2): pp. 41–47.
270. Yang, L., *Development of in-field transportation robot vehicle using multiple sensors*. In *Department of Environment Resources, Graduate school of Agriculture*. 2013, Hokkaido University. p. 138.
271. Barawid Jr, O.C., *Development of electronic utility robot vehicle*. In *Department of Environment Resources, Graduate school of Agriculture*. 2011, Hokkaido University. p. 170.
272. Ospina, R. and N. Noguchi, Determination of tire dynamic properties: Application to an agricultural vehicle. *Engineering in Agriculture, Environment and Food*, 2016. **9**(1): pp. 123–130.
273. Ospina, R. and N. Noguchi, Alternative method to model an agricultural vehicle's tire parameters. *Engineering in Agriculture, Environment and Food*, 2018. **11**(1): pp. 9–18.
274. Yang, L., N. Noguchi, and R. Takai, Development and application of a wheel-type robot tractor. *Engineering in Agriculture, Environment and Food*, 2016. **9**(2): pp. 131–140.
275. Yang, L. and N. Noguchi, An active safety system using two laser scanners for a robot tractor. *IFAC Proceedings Volumes*, 2014. **47**(3): pp. 11577–11582.
276. Noguchi, N., Current status and future prospects for vehicle robotics on agriculture. *Journal of the Japan Society for Precision Engineering*, 2015. **81**(1): pp. 22–25.
277. Zhang, C., L. Yang, and N. Noguchi, *Development of a human-driven tractor following a robot system*. In *Proceedings of the 19th World Congress, The International Federation of Automatic Control*. 2014, IFAC Proceeding Cape Town, South Africa.
278. Zhang, C., et al., Development of robot tractor associating with human-drive tractor for farm work. *IFAC Proceedings Volumes*, 2013. **46**(4): pp. 83–88.
279. Zhang, C., L. Yang, and N. Noguchi, Development of a robot tractor controlled by a human-driven tractor system. *Engineering in Agriculture, Environment and Food*, 2015. **8**(1): pp. 7–12.
280. Zhang, C., *Development of a leader-follower system for farm use*. In *Department of Environment Resources, Graduate school of Agriculture*. 2014, Hokkaido University.

281. Yang, L., *Development of a robot tractor implemented an omni-directional safety system*. In *Department of Environment Resources, Graduate school of Agriculture*. 2013, Hokkaido University. p. 133.

282. Zhang, Z., et al., Development of a robot combine harvester for wheat and paddy harvesting. *IFAC Proceedings Volumes*, 2013. **46**(4): pp. 45–48.

283. Zhang, Z., *Development of a robot combine harvester based on GNSS*. In *Department of Environment Resources, Graduate school of Agriculture*. 2014, Hokkaido University.

284. Zhang, Z., et al., Optimization of steering control parameters based on a combine harvester's kinematic model. *Engineering in Agriculture, Environment and Food*, 2014. **7**(2): pp. 91–96.

285. Rahman, M., K. Ishii, and N. Noguchi, Study on tracked combine harvester dynamic model for automated navigation purposes. *Advances in Robotics & Automation*, 2017. **6**(3).

286. Rahman, M., *Studies on tracked dynamic model and optimum harvesting area for path planning of robot combine harvester*. In *Department of Environment Resources, Graduate school of Agriculture*. 2018, Hokkaido University.

287. Choi, J., et al., Development of a laser scanner-based navigation system for a combine harvester. *Engineering in Agriculture, Environment and Food*, 2014. **7**(1): pp. 7–13.

288. Teng, Z., et al., Development of uncut crop edge detection system based on laser rangefinder for combine harvesters. *International Journal of Agricultural & Biological Engineering*, 2016. **9**(2): pp. 21–28.

289. Fan, Z., *Development of a real-time quality sensor for wheat on a combine harvester*. In *Department of Environment Resources, Graduate school of Agriculture*. 2015, Hokkaido University.

290. Choi, J., X. Yin, and N. Noguchi, Development of a Laser Scanner-based Navigation System for a Combine Harvester. *IFAC Proceedings Volumes*, 2013. **46**(18): pp. 103–108.

291. Liu, Y. and N. Noguchi, Development of an unmanned surface vehicle for autonomous navigation in a paddy field. *Engineering in Agriculture, Environment and Food*, 2016. **9**(1): pp. 21–26.

292. Liu, Y., N. Noguchi, and A. Roshanianfard, Simulation and test of an agricultural unmanned airboat maneuverability model. *International Journal of Agricultural and Biological Engineering*, 2017. **10**(1): pp. 88–96.

293. Liu, Y., et al., *Wind direction-based path planning for an agricultural unmanned airboat navigation*. In *2016 IEEE/SICE International Symposium on System Integration (SII)*. IEEE, 2016.

294. Okada, A., *Study on automatic steering control of rice transplanter by satellite positioning system (In Japanese)*. In 2017, Hokkaido University.

295. Kawahito, H., *Study on automatic steering of rice transplanter by detection of traveling marker using image processing (In Japanese)*. In *Agriculture Department*. 2016, Hokkaido University.

296. Wada, *Research on rudder automatic steering for labor saving in rice transplanter*. In *Agriculture Department*. 2017, Hokkaido University.

297. Wang, H. and N. Noguchi, *Autonomous maneuvers of a robotic tractor for farming*. In *2016 IEEE/SICE International Symposium on System Integration (SII)*. 2016. IEEE.

298. Zhang, C. and N. Noguchi, *Development of leader-follower system for field work*. In *2015 IEEE/SICE International Symposium on System Integration (SII)*. 2015. IEEE.

299. Zhang, C., N. Noguchi, and L. Yang, Leader–follower system using two robot tractors to improve work efficiency. *Computers and Electronics in Agriculture*, 2016. **121**: pp. 269–281.

300. Noguchi, N. and O.C. Barawid, Robot farming system using multiple robot tractors in Japan agriculture. *IFAC Proceedings Volumes*, 2011. **44**(1): pp. 633–637.

301. Zhang, C. and N. Noguchi, Development of a multi-robot tractor system for agriculture field work. *Computers and Electronics in Agriculture*, 2017. **142**: pp. 79–90.

302. Zhang, C., *Development of a multi-robot tractor system for farm work*. In *Department of Environment Resources, Graduate School of Agriculture*. 2017, Hokkaido University.

303. Roshanianfard, A. and N. Noguchi, *Development of a heavyweight crop robotic harvesting system (HCRH)*. In *2017 The 3rd International Conference on Control, Automation and Robotics*. 2017, IEEE.

304. Roshanianfard, A. and N. Noguchi, *Development of robotic harvesting system for heavy-weight crops*. In *11th Seminar of ASIJ –Academic Society of Iranians in Japan*. 2017. Tokyo, Japan.

305. Roshanianfard, A. and N. Noguchi, Development of a 5DOF robotic arm (RAVebots-1) applied to heavy products harvesting. *IFAC-PapersOnLine*, 2016. **49**(16): pp. 155–160.

306. Kamata, T., A. Roshanianfard, and N. Noguchi, Heavy-weight crop harvesting robot – controlling algorithm. *IFAC-PapersOnLine*, 2018. **51**(17): pp. 244–249.

307. Roshanianfard, A. and N. Noguchi, Kinematics analysis and simulation of a 5DOF articulated robotic arm applied to heavy products harvesting. *Tarim Bilimleri Dergisi-Journal of Agricultural Sciences*, 2018. **24**(1): pp. 91–104.

308. Roshanianfard, A., N. Noguchi, and T. Kamata, Design and performance of a robotic arm for farm use. *International Journal of Agricultural and Biological Engineering (IJABE)*, 2019. **12**(1): pp. 146–158.

309. Roshanianfard, A. and N. Noguchi, Characterization of pumpkin for a harvesting robot. *IFAC-PapersOnLine*, 2018. **51**(17): pp. 23–30.

310. Roshanianfard, A. and N. Noguchi, *Designing of pumpkin harvester robotic end-effector*. In *2017 The 3rd International Conference on Control, Automation and Robotics (ICCAR 2017)*. 2017, Nagoya, Japan: IEEE.

311. Roshanianfard, A., T. Kamata, and N. Noguchi, Performance evaluation of harvesting robot for heavy-weight crops. *IFAC-PapersOnLine*, 2018. **51**(17): pp. 332–338.

312. Bac, C.W., et al., Harvesting robots for high-value crops: State-of-the-art review and challenges ahead. *Journal of Field Robotics*, 2014. **31**(6): pp. 888–911.

Index

Printed in the United States
by Baker & Taylor Publisher Services